The Milky Way

The Milky Way

An Insider's Guide

· · • • · ·

William H. Waller

Princeton University Press

Princeton and Oxford

press.princeton.edu

A composite photo of the summer Milky Way appearing
to stand on end over a deserted road in Texas
by Larry Landolfi. © Photo Researchers, Inc.

ISBN 978-0-691-12224-3
Library of Congress Control Number: 2013930365

British Library Cataloging-in-Publication Data is available

This book has been composed in Sabon LT Std text
with Helvetica Neue LT Std Display

Printed on acid-free paper ∞

Printed in the United States of America

1 3 5 7 9 10 8 6 4 2

CONTENTS

· · ● ● ● · ·

PREFACE

• • • • • •

I FIRST BECAME ENTHRALLED with the Milky Way as a boy of ten. My family had moved two years prior—from the light-polluted urban landscape of Mount Vernon, New York to the quaint sea-side town of Rockport, Massachusetts. In those two years, I slowly became acclimated to my human and physical surroundings and began to look up. Like any other child, I first noticed the Moon and its changing phases, followed by the brighter planets and their odd meanderings among the stars. But one clear fall night in 1962, I saw something entirely new and different. Looking straight up from my driveway, I beheld a gauzy wisp of light beyond the tree-tops that defied explanation. My guess now is that I was looking toward the constellations of Cygnus and Cassiopeia, where the autumnal Milky Way is highest and most prominent. But back then, I knew nothing of these things. That ghostly light evoked feelings of mystery and delight, as if I had chanced upon a hidden treasure.

My parents had noted my growing interest in the night sky, and so for Christmas, they gave me a small refracting telescope. That night, I padded out onto the crusty snow in my robe and galoshes and aimed my new spyglass skyward. Almost immediately, I learned how difficult it is to keep the telescope properly pointed at the Moon or at a planet like Jupiter. Bright stars were just as chal-lenging to fix upon, but when I got one sighted, the view was amazing. Bloated and colorful, each star took on a fantastic per-sona. It didn't take long for me to discover that these stellar spec-tacles were the artifacts of my poor attempts to focus the telescope. Chided but undeterred, I continued to improve my observing technique by first focusing on the Moon and then on a bright planet. Once I had refined the focus, I was astounded—and initially disappointed—to find that the brightest stars appeared stubbornly unmagnified. I was old enough to make the connection. If the stars are like the Sun, then they must be incredibly far away to look so small. In one frigid night of crude observations from my backyard, I had discovered just how vast the universe must be.

Turning my telescopic attention to the Milky Way, I saw that the diffuse light resolved into individual stars. Just like Galileo Galilei had done 352 years before me, I could see for myself that the heavens were ruled by stars. From then on, the Milky Way has been my companion and confidant. I continue to be star-struck whenever I get the chance to see the Milky Way in all its glory—from a sandy beach on Cape Cod or a mountaintop observatory in Chile. I am also grateful to have participated in (and benefited from) several astronomical surveys at optical, infrared, and radio wavelengths that have culminated in wondrous maps of the Milky Way. These and many other multi-wavelength vistas have enabled astronomers to piece together some amazing stories about the structure, dynamics, origin, and evolution of our home galaxy. This book is an attempt to tell these stories in a coherent and compelling manner.

In some ways, the title is a bit of a cruel joke. We know today that we are inhabitants of an incredibly vast galaxy. We also know that we are essentially doomed to remain "insiders"—stuck in a tiny part of this realm, never to see its full expanse from afar. Given these circumstances, we can do three things. We can marvel at and learn from the many amazing stellar and nebular objects that can be perceived from our particular vantage point. We can translate these observations into a fully functioning three-dimensional model of the Milky Way Galaxy. And we can compare this model with observations of other galaxies, both nearby and far away.

The last book to take such an approach was *The Milky Way* by Bart and Priscilla Bok (Harvard University Press). That book went through five editions, the final edition appearing in 1981—two years before Bart Bok's death. Given this connection, and the personal friendship that I enjoyed with Bart, I dedicate this book to his and Priscilla's memory.

Since the Boks' last account in 1981, astonishing changes have occurred in our understanding of the Milky Way. New surveys at radio, infrared, optical, and X-ray wavelengths reveal a galaxy that is as rich, complex, and dynamic as one could ever imagine. Indeed, the overall specifications of the Milky Way have undergone radical revision. From the inside out, these include a strangely quiet supermassive black hole in the nucleus, a humongous bulge of stars in the central 3,000 light-years, an elongated central "bar" of stars that extends halfway out to the Sun's orbit some 27,000 light-years from the nucleus, a ringlike concentration of gas clouds and newborn stars just beyond the stellar bar, spiral arms of gas and young stars unfolding from the ring to well beyond the Sun's orbit, a stylish warping of the outer disk evident in both the gas and stars, a spheroidal halo of ancient stars and globular star clusters extending out to a radius of 100,000 light-years, and stellar streams streaking through the halo that are thought to be tidal debris from in-falling dwarf galaxies. The halo itself is thought to harbor fantastic amounts of dark matter—amounting to more than 85 percent of the Galaxy's overall mass.

The exquisitely complex relations among these various components reveal the Galaxy as resembling a living, breathing organism—one that emerged from the chaos of the Hot Big Bang some 12 billion years ago, and that is still very much alive with the pyrotechnics of star birth and star death. Along the way, the Milky Way has hosted the spawning of life on the moist surface of one particular planet, and may be fostering similar biological experiments elsewhere. In the last chapters of this book, I explore the biochemical character of the Milky Way and how scientists are looking for the telltale signs of life beyond the Solar System. Recent discoveries of exoplanets pretty much everywhere we have looked have dramatically increased the prospects for finding planets with chemistries favorable to life. Indeed, the potential for phenomenal progress in this emerging field of astrobiology has never been greater.

I end the book with a *Galactic Manifesto* that presents the case for humankind to take on our rightful roles and responsibilities as communicative citizens of the Milky Way. Although we may never make direct contact with other self-conscious life-forms beyond Earth, I contend that we have the moral obligation to conduct our lives as if that contact were imminent. In this *Galactic Manifesto*, I see our place in the cosmos as true insiders—beneficiaries of exquisite stellar and nebular processes that have transpired over 12 billion years of galactic history. Today, in deploying planetary robots, interplanetary probes, sensitive telescopes, and powerful transmitters, we have already begun to take our first steps as fully vested members of the Milky Way. We should begin to think and act accordingly.

In writing this book, I received the kind attention of many astronomers, science historians, friends, and family members. I gratefully acknowledge the insights and materials provided by Lori Allen, Bruce Balick, Robert Benjamin, Leo Blitz, Adam Block, Hale Bradt, Tom Dame, Edna DeVore, Paul Goldsmith, Allen Hirshfeld, Paul Hodge, John Huchra, Alaa Ibrahim, James Kaler, Dan Klepinger, Andrew Knoll, Edwin C. Krupp, Charlie Lada, Jay Lockman, Oleg Malkov, Massimo Marengo, Axel Mellinger, Mark Reid, Dimitar Sasselov, Norbert Schulz, Pat Slane, Jonathan Slavin, and Karen Wade. Of course, none of these folk are responsible for any errors of fact or interpretation that might still lurk within these pages. I am grateful to Nanette Benoit, Dan Lampert, and Leigh Slingluff, who crafted many of the figures that adorn and inform this book. Leigh in particular worked many miracles in making the visuals shine. I am especially grateful to Joan Paille, who diligently secured permissions to use figures from other sources.

I have also benefited from the resources and guidance provided by library staff at the Rockport Public Library in Rockport, Massachusetts, Corning Museum of Glass in Corning, New York, Harvard-Smithsonian Center for Astrophysics and Harvard Origins of Life Initiative in Cambridge, Massachusetts, and Tufts University in Medford, Massachusetts. At Tufts University, Provost Jamshed Bharucha, Dean Robert Sternberg, Physics and Astronomy Department Chair Bill Oliver, and many of my former astronomy students provided much appreciated support during the book's early gestation. Ingrid Gnerlich at Princeton University Press helped carry the book forward through her constructive prodding, encouragement, and sage advice.

Finally, I never would have been able to pursue such a far-reaching but self-centered endeavor were it not for the loving support of my spouse Sandra Paille, our delightful progeny—Julian Waller and Renée Waller, my mother Pat Waller, sister Sue Waller, uncle Al Waller, and aunt Kim Waller. May this book in some way reward them all (both living and deceased) for their unflagging encouragement of my scientific and literary efforts.

To keep abreast of the latest discoveries, and to delve more deeply into other topics pertaining to the Milky Way, the reader is invited to peruse the following website (http://sites.google.com/site/thegalacticinquirer) and to contact me directly with any questions or comments at williamhwaller@gmail.com.

William H. Waller

Rockport, Massachusetts
June 2012

The Milky Way

CHAPTER 1

• • • • • •

FIRST IMPRESSIONS

You can observe a lot by watching.
—Yogi Berra (1925–)

IMAGINE YOURSELF ON A MAGIC CARPET, levitating away from Earth on a voyage into deep space. As you begin your ascent, you can see ever enlarging vistas of land and sea beneath you. Very soon, the terrestrial horizon begins to curve and fall away. Your initial concept of a straight horizon that segregates Earth from sky has become nonsensical. Instead, you see your home orb shrinking ever smaller, and the starry sky enlarging to fill the expanse. As your flying tapestry propels you beyond the inner Solar System, your view of the Sun also begins to take up less and less of the sky. Somewhere, way beyond the orbit of Pluto, you look back to see that the planets have all but disappeared into the inky darkness, with the Sun now only one of many bright stars in the firmament. You are now entering the realm of interstellar space.

Comfortably perched upon your magical mat, you can see stars above you and below you. You are completely enfolded in starlight. The familiar constellations are still there, including Ursa Major the Big Bear, Orion the Hunter, and Leo the Lion. But there is something else that you can now see as never before. A diaphanous band of eerie light completely encircles you like a hazy ring of muted fire. This irregular skein of milky luminescence appears to connect with you somehow, and indeed, your initial impression is correct. Having removed yourself from the obstructing Earth and blinding Sun, you have situated yourself in and amongst the

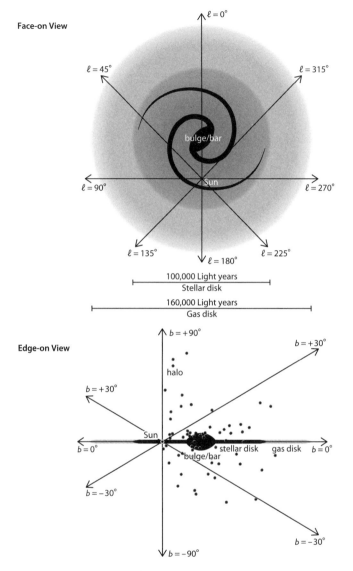

FIGURE 1.1. Cartoon schematic of the Milky Way Galaxy as seen from well beyond its extremities, showing both face-on and edge-on perspectives. We live inside the Milky Way, and so cannot directly view our home galaxy from afar, but can construct models based on what we have observed from within. (*Top*) Face-on perspective of the Milky Way featuring the central bulge/bar of stars, stellar and gaseous disk, and spiral arms in the disk that trace recent star-forming activity. Lines of constant galactic longitude demarcate the view along the Milky Way as seen from our particular location inside the galactic disk (see plates 2–5 and figure 1.2). (*Bottom*) Edge-on perspective of the Milky Way featuring the central bulge/bar, thin stellar and gaseous disk, and halo that contains globular star clusters (shown) and massive amounts of dark matter (not shown). Lines of constant galactic latitude delineate the views in plates 2–5 and figure 1.2. [Courtesy of W. H. Waller and D. Lampert]

stars that comprise the flattened disk of the Milky Way Galaxy (see figure 1.1). Taking out your trusty binoculars, you confirm that the encircling haze consists of stars upon stars extending away into the vastness of space. You are awash in the Milky Way, your cosmic home.

Mapping the Milky Way

Though mostly earthbound, astronomers have been able to achieve the same all-sky view that your carpet-borne fantasy has suggested. By taking pictures from both the northern and southern hemispheres of Earth and by stitching these pictures into mosaics, they have created all-sky maps that reveal the entirety of the Milky Way (see plate 1). Projecting these all-sky views onto flat pieces of paper presents significant problems, however, as major distortions are inevitable. The best we can do is to project the Milky Way itself as a single undistorted band that runs along the "equator" of these maps, while letting other parts of the sky become distorted. Doing so leads to a natural system of "galactic coordinates" whose longitude is measured along the galactic equator and whose latitude runs perpendicular to the equator positively toward the North Galactic Pole and negatively toward the South Galactic Pole. Ground zero for the galactic coordinate system is located in the constellation of Sagittarius the Archer, where the Milky Way appears brightest and most extended in latitude. Its exact position has changed over the years, as astronomers have learned more about the true galactic center and its precise location in the sky.

The Inner Galaxy

Looking closer, we can see that each segment of the Milky Way has its own unique features. Between longitudes of 310° (–50°) → 360° (0°) → +50°, our attention is directed toward the inner Galaxy (see plate 2 and figure 1.2). This is where the Milky Way is brightest—and for good reason—as we are viewing the most pop-

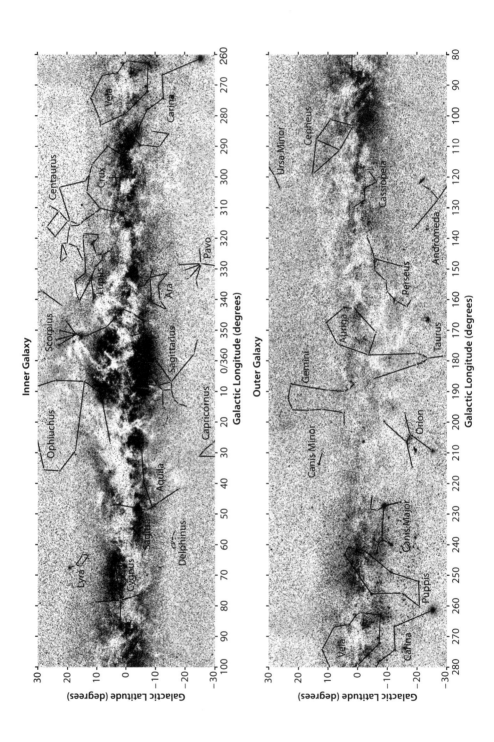

Inner Galaxy

Galactic Longitude (degrees)

Galactic Latitude (degrees)

Vela, Carina, Centaurus, Crux, Ophiuchus, Scorpius, Lupus, Ara, Pavo, Sagittarius, Capricornus, Aquila, Sagitta, Delphinus, Lyra, Cygnus

Outer Galaxy

Galactic Longitude (degrees)

Galactic Latitude (degrees)

Ursa Minor, Cepheus, Cassiopeia, Perseus, Andromeda, Auriga, Taurus, Gemini, Orion, Canis Minor, Canis Major, Puppis, Vela, Carina

ulated part of the Galaxy. From our vantage point, we can see parts of the central bulge extending above and below the disk component. In long-exposure images, we can also spot several roseate nebulae and bluish star clusters, hallmarks of ongoing star formation. The brightest of these fuzzy patches have been given names that evoke their visual appearance through small telescopes. The Lagoon Nebula, Eagle Nebula, Omega Nebula, and Flying Duck Cluster are just a few of the marvels that can be found toward the inner galaxy.

Most of these celestial objects are located in the Sagittarius spiral arm, the nearest arm to us in this part of the Milky Way. Much is hidden from view, however, as thick tendrils of dust obscure the light from the innermost disk and bulge. Indeed, we can peer into the inner galaxy only about 5,000 light-years before the intervening clouds of dust impede any further visible reconnaissance. That gets us only one-fifth of the way toward the galactic center, which lies a full 27,000 light-years from us. Some of the dusty obscuration can be viewed naked-eye as rifts of darkness that cleave the Milky Way into strange shapes. The Great Rift is most notable, extending from the constellation of Sagittarius through Serpens the Serpent, Scutum the Shield, and Aquila the Eagle. Similar but smaller blobs of darkness are apparent on the other side of Sagittarius in the constellations of Ophiuchus the Serpent Bearer, Scorpius the Scorpion, Norma the Square, and Circinus the Compasses. In the following chapters, we will learn a lot more about these murky regions, where new stars are being born.

FIGURE 1.2. (*Top*) Negative rendering of the visible Milky Way (where stars appear black) looking toward the inner galaxy and galactic center. Galactic longitude increases from right to left—from 260° through 360° (0°, the galactic center) to 100°. Galactic latitude ranges from bottom to top, from -30° through 0° (the galactic midplane) to +30°. Major constellations are shown. (*Bottom*) The visible Milky Way looking toward the outer galaxy and galactic anti-center. Galactic longitude increases from right to left, from 80° through 180° (the galactic anti-center) to 280°. [Images courtesy of A. Mellinger (Central Michigan University); see http://home.arcor-online.de/axel.mellinger/mwpan_old.html]

Looking Downstream

Between the longitudes of 40° → 90° → 140°, we are looking in the general direction of our Solar System's orbital motion around the galactic center (see plate 3 and figure 1.2). Our sight lines intersect the constellations of Sagitta the Arrow, Vulpecula the Fox, Cygnus the Swan, Cepheus the King, and Cassiopeia the Queen. The Orion and Perseus spiral arms make their presence known in the form of myriad dark clouds and fluorescing nebulae. One of these nebulae—the North America Nebula in Cygnus—is large enough to be seen naked-eye under optimal dark-sky conditions. Like other nebulae of its kind, the North America Nebula is being made to fluoresce by the intense ultraviolet light from newborn hot stars in its immediate vicinity. There is another nebula in Cygnus that is worth noting, for it traces the shock waves that have been rippling through interstellar space since the explosion of a massive star some 10,000 years ago. The Cygnus Loop is a favorite target for amateur astronomers equipped with telescopes of 10-inch to 20-inch diameter apertures.

The Outer Galaxy

Beyond 130° longitude and extending through 180° to 230°, we are looking away from the galactic center and towards the galactic anti-center (see plate 4 and figure 1.2). If we could regard ourselves as living in the suburbs of the Milky Way Galaxy, then we would be looking away from the Big City and toward the rural woodlands beyond. The intensity of diffuse starlight is considerably less along this stretch of the Milky Way, but there are plenty of fascinating sources to catch our eye.

Beginning with Perseus the Rescuer of Andromeda, we find the famous Double Cluster, also known as h and chi Persei. Through binoculars, a glittering raiment of young blue stars can be perceived. Perseus adjoins Auriga the Charioteer and Gemini the Twins, where binoculars will reveal several clusters of stars. Like the Double Cluster, these clusters are known as "open" clusters, owing to their loosely organized appearance.

From Auriga, we pass through Taurus the Bull—home of the Taurus Molecular Cloud. At a distance of 400 light-years, the Taurus Molecular Cloud is one of the nearest stellar nurseries, where dust-enshrouded protostars are incubating. Visible-light photographs delineate this cloud of gas and dust as a rivulet of darkness, silhouetted against the background stars. The constellation of Taurus also hosts the Pleiades and Hyades star clusters, the most prominent clusters in the naked-eye sky. The Pleiades brightly shine with the bluish light of hot young stars. The Hyades star cluster is considerably older and so shines with the mellower yellow and orange colors characteristic of its longer-lived stars.

Below Taurus, Orion the Hunter takes aim at the bull with his shield and club. The Orion constellation contains an impressive assortment of hot stars and glowing clouds—testimony to the giant molecular cloud in this part of the sky. Situated about 1,500 light-years from us, the Orion Molecular Cloud is only visible to our eyes where newborn stars are illuminating its sundry gaseous surfaces. The Orion Nebula and Horsehead Nebula are two notable instances, where parts of the larger cloud have been lit up like storm clouds illuminated by fireworks. Similar scenes of recent star-forming activity are being played in nearby Monoceros the Unicorn, home of the Rosette Nebula and Cone Nebula. Like the Orion and Horsehead nebulae, the Rosette and Cone nebulae feature clusters of hot stars that are irradiating, excavating, and fluorescing the nebulosity that surrounds them. Good binoculars and a clear dark night are just barely sufficient for one to detect the Rosette Nebula as something more than a tight grouping of stars. The Cone Nebula requires more powerful telescopic assistance, and is best detected in long-exposure images.

Looking Upstream

The longitudes of $220° \rightarrow 270° \rightarrow 320°$ take us in the direction opposite to the Solar System's orbital motion around the galactic center (see plate 5 and figure 1.2). The constellations of Canis Majoris the Big Dog, Puppis the Stern, Vela the Sails, Carina the Keel, Centaurus the Centaur, and Crux the Cross complete our inventory of major star patterns along the Milky Way. Much of this sector can only be viewed from the Earth's southern latitudes—a fine excuse to visit South America, South Africa, or Australia! Once again, our gaze extends along a few of the Galaxy's spiral arms, most notably the Perseus and Carina-Sagittarius arms. Among these large-scale (and still poorly defined) spiral features, three naked-eye objects stand out. The Carina Nebula is the largest and brightest emission nebula in the sky. It spans more than two full moons and outshines the Orion Nebula in absolute terms by a factor of 50. Contrasting with the Carina Nebula, the Coal Sack obscures rather than illuminates. This patch of inky darkness appears absolutely opaque against the ghostly backdrop of more distant stars. Long photographic exposures, however, show that the Coal Sack is clumpy—with several stars visible within its dark ramparts. Abutting the Coal Sack, the Southern Cross (Crux) blazes forth with its distinct pattern of five bright stars. The flags of Australia, New Zealand, Papua New Guinea, and Samoa are each graced with graphic renderings of the Southern Cross. Our 360° tour of the Milky Way ends with Alpha Centauri—the nearest stellar system to us. At a distance of only 4.2–4.4 light-years, the three stars that make up the Alpha Centauri system share our perspective of the larger Milky Way. If we could converse with any planetary inhabitants that happened to be in orbit around these stellar realms, we would have no difficulty chatting about the Milky Way.

Above and Below

Of course, not all of the Milky Way Galaxy is confined to the narrow band that we call the "Milky Way." All of the stars that are visible in the sky belong to our home galaxy. Indeed, the only naked-eye objects that are *not* part of the Milky Way Galaxy are the Andromeda Galaxy that is visible from northern latitudes and the Large and Small Magellanic Clouds that can be seen from southern latitudes. As seen by us, many of the Sun's most prominent stellar neighbors are arrayed well above and below the Milky Way. Although these stars are still part of the Galaxy's disk component, their close proximity to us allows for large (apparent) deviations from the disky norm.

Other galactic objects are located far from the Milky Way because they are in fact situated far from the disk—in a part of the Galaxy that is called the halo. Globular star clusters are most notable for populating the halo component. These dense stellar swarms contain the oldest stars in the Galaxy. Their origin some 12 billion years ago continues to puzzle astrophysicists, as they implicate a galaxy that was very different in the distant past. Did today's Milky Way Galaxy evolve from a humongous cloud that collapsed under its own weight to form a rapidly rotating disk with leftover globular star clusters in the halo? Or did the Milky Way Galaxy get pieced together from a flurry of merging dwarf galaxies—where the globular star clusters represent the undigested cores of these dwarf systems? Many questions remain regarding the structure, origin, and evolution of our galactic home. To answer them, astronomers are drawing on multi-wavelength observations of every component that they can detect. Some of these mind-expanding observations are described in chapter 3. But first, an abridged history of our evolving galactic perceptions is in order.

CHAPTER 2

· · ● ● · ·

HISTORIC PERCEPTIONS

Wisdom begins with wonder.
—Socrates (470–399 BCE)

HUMANS HAVE MARVELED at the Milky Way for as long as they
have roamed the surface of Earth. Indeed, the ancient hominids of
the African savannah enjoyed views of the Milky Way that were
far superior to those experienced by most of us in the modern
world. Blessed with dry, clear skies that were free of light pollu-
tion, these earliest sky gazers and their later descendents have born
witness to the Milky Way for hundreds of millennia. To ponder the
Milky Way today is to share our cosmic wonder with these primeval astronomers, and in doing so, to commune with our human
heritage.

Archaic Visions

The first records of the night sky come from megalithic monuments that have withstood the ravages of time. These include Newgrange and other astronomically aligned and decorated "passage
tombs" in Ireland that date back to before 3200 BCE, Stonehenge
in southern England (circa 3100–1600 BCE), and the great pyramids of ancient Egypt (circa 2500–1000 BCE). Thanks to the
translational Rosetta Stone, we know the most about the ancient
Egyptian records of the sky. These include bas-relief sculptures,

paintings on various surfaces, and associated hieroglyphics that depict the sky goddess Nut (see figure 2.1). As Nut represented the vault of heaven, she was often portrayed with all of the celestial accoutrements befitting her role. She was thought to be responsible for swallowing the Sun at sunset and for giving birth to it at sunrise. The same goes for the Moon, the constellation of Orion, and other bright objects in the night sky. Her role as receiver, reviver, and protector of the dead made Nut one of the most popular of the great Egyptian goddesses. Her image was frequently painted on the inside bottom or lid of coffins. In one such coffin is written this prayer—"O my mother Nut, stretch yourself over me, that I may be placed among the imperishable stars, which are in you, and that I may not die."

The specific relationship between Nut and the Milky Way is less clear. As a mother goddess, she could have expressed the Milky Way as a celestial river of divine milk. Yet precious few words in the Pyramidal Texts and other writings mention Nut and the Milky Way together. Several Egyptian scholars directly identify the lanky form of Nut with the Milky Way, but this claim remains controversial.

One of the first explicit references to the Milky Way comes from the Acropolis at Mycenae (located in the south-central part of present-day Greece). A gold signet ring of Bronze Age vintage (circa 1400 BCE) portrays an abundance of allusions to the Milky Way—including buxom goddesses (or priestesses) brandishing poppy plants whose milky sap was even then known to produce narcoleptic languor. Above the heads of these magical maidens and beneath obvious references to the Sun and Moon are two parallel undulating lines that are thought to represent the Milky Way itself (see figure 2.1).

Another thousand years later, the Ionian Greek scholar Democritus would write of the Milky Way as "a luster of small stars very close together." This prescient interpretation was consistent with the atomist view of matter that Democritus shared with his fellow Epicureans. Yet another two thousand years would pass, before his supposition was confirmed by Galileo's telescopic observations.

FIGURE 2.1. Ancient Near Eastern depictions of the night sky. (*Top*) The Egyptian sky goddess Nut arching over the Earth god Geb and atmospheric god Shu, c. 2000 BCE. (*Bottom*) Imprint of Mycenaen signet ring with symbols for the Sun, Moon, and a rippling Milky Way, c. 1400 BCE. [(*Top*) Illustration adapted from multiple sources, with reference to *The Great Goddesses of Egypt* by B. S. Lesko. (*Bottom*) Adapted from a golden signet ring from Mycenae, c. 1500 BCE, curated at the National Archaeological Museum, Athens. Discussed by E. C. Krupp in "Spilled Milk," *Griffith Observer* 57, no. 12 (1993), with reference to *The Minoan-Mycenaean Religion and Its Survival in Greek Religion* by M. P. Nilsson. Lund: C.W.K. Gleerup, 1949]

Meanwhile, the dominant Greco-Roman interpretation of the Milky Way was as a river of milk. According to legend, the creation of this river began with the goddess Hera, wife of Zeus, whose breast milk was known to confer immortality on anyone who suckled upon her. Hera suffered with jealousy from the many infidelities that Zeus pursued, and the illegitimate children that resulted from his trysts. These included Hermes, son of the nymph Maia, and Heracles, son of the mortal Alcmene. While Hera was sleeping, Hermes placed the half-mortal Heracles to her breasts with the intent of conferring immortality upon the infant. Upon awaking, Hera was appalled to find this illegitimate spawn nursing at her nipple and pushed him away. The spilt milk sprayed into the sky, thus forming the Milky Way. This melodramatic myth has been depicted by several artists, including Tintoretto and Rubens. Although today's astronomers do not concur with this physical interpretation, they have borrowed from the ancient Greeks the word "galaxy" which originated from "Galaxias Kuklos" meaning "Milky Circle."

Beyond the Mediterranean basin, other ancient cultures have perceived and represented the Milky Way in their own unique ways. Chinese poets dating back to the Jin Dynasty (256–420 CE) wrote hundreds of verses recounting the story of Niu Lang and Zhi Nu—two lovers separated by the "Silver River" of the Milky Way. According to these poems, the bright star Vega in the constellation of Lyra the Harp, located east of the Milky Way, is Zhi Nu, the weaving maiden, and the star Altair in the constellation of Aquila the Eagle, to the west of the Milky Way, is her husband Niu Lang, the cowherd. This star-crossed couple was placed in the sky after neglecting their duties and so irking Zhi Nu's mother, the Queen of Heaven. Once a year, however, they were given a reprieve from their segregated exiles. On the seventh day of the seventh lunar month (which typically falls in August), Niu Lang and Zhi Nu would meet on a bridge of magpies that spanned the divisive Milky Way. One of the most famous poems about this legend was written by Qin Guan in the Song Dynasty (960–1279 CE). Here is a translation by Kylie Hsu, professor of Chinese Studies and Linguistics at California State University, Los Angeles.

FAIRY OF THE MAGPIE BRIDGE

> Among the beautiful clouds,
> Over the heavenly river,
> Crosses the weaving maiden.
> A night of rendezvous,
> Across the autumn sky,
> Surpasses joy on earth.
> Moments of tender love and dream,
> So sad to leave the magpie bridge.
> Eternal love between us two,
> Shall withstand the time apart.

This romantic interlude amidst the Milky Way continues to be celebrated in China. Girls prepare offerings to Zhi Nu, the exiled maiden, and pray to find good husbands. In the evening, people gaze upon the stars with earthly yearning in their hearts.

In aboriginal Australia, the Milky Way played a key role as an arbiter of Creation. It was known as the great serpent Wallanganda. The Earth was a serpent too, called Ungut. Together, Wallanganda and Ungut gave birth to the Creation by dreaming all the creatures that live on Earth—including the aboriginal spirit ancestors and the magical Wandjina who bring both rain and fertility. Ancient rock paintings of Wallanganda, Ungut, and Wandjina have been found in the Kimberly outback of northwest Australia. Though continuously repainted by contemporary aborigines, this rock art is thought to have originated more than 17,000 years ago.

In the Americas, indigenous astronomers have created many stories and artifacts that express their fascination with the Milky Way. The Desana Indians of the Amazon rainforest saw in the Milky Way two intertwining snakes—one a male rainbow boa, the other a female anaconda. The braided nature of this imagery can be perceived in the plaited patches of light and darkness that undulate along the Milky Way. For the Mexican Aztecs, a similar duo of serpents plays a central role in their story of creation. The famous Aztec Sun Stone portrays elements of this myth, with two dragons encircling its perimeter and the creation gods Quetzalcoatl and Texcatlipoca emerging from the serpents' mouths (see figure 2.2). According to some ethnologists, these serpents repre-

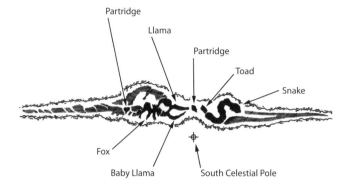

FIGURE 2.2. Indigenous American depictions of astronomical themes. (*Top*) "Dark" constellations along the Milky Way envisaged by the Quechua Indians of the Peruvian Andes. (*Bottom*) Aztec Sun Stone with astronomical motifs including the central sun god, star patterns on the outside, and two serpents along the stone's perimeter. These serpents have been interpreted by some ethnologists as representing the Milky Way, but may have other meanings. [(*Top*) Drawing by J. Bieniaz (Griffith Observatory). Courtesy of E. Krupp (Griffith Observatory). (*Bottom*) c. 1479 CE, curated at the National Museum of Anthropology in Mexico City. Discussed by B. C. Brundage in *The Phoenix of the Western World, Quetzalcoatl and the Sky Religion*, Norman: University of Oklahoma Press, 1982]

sent the Milky Way—an intriguing claim that remains speculative, however.

Quechua Indians of the Peruvian Andes focused on the dark patches of the Milky Way, naming them after local animals. Their highly structured view of the Milky Way is amazingly true to what we can actually see on clear, moonless nights. For example, the Coal Sack that lies next to the brilliant Southern Cross is clearly represented as Yutu—the Partridge (see figure 2.2).

The Maya of present-day Guatemala, Belize, and Mexico explained the Milky Way as a "World Tree." Rooted in the underworld and branching up into the sky, the World Tree provided a means of transport from one realm to the other. The Indians of California called the Milky Way the "sky's backbone," "ghost trail," and "pathway of the spirits," depending on the particular tribe and locale. Their representational artifacts include rock carvings, sand paintings, ceremonial "ground displays," and headbands made of eagle down.

Many of these diverse cultures held remarkably similar perceptions of the Milky Way as a welcoming home for the dead or as a celestial pathway for the gods and deceased heroes. As archeoastronomer Edwin Krupp puts it, "Across cultural boundaries, the Milky Way retains its abstract, philosophical, and religious dimensions. Its imagery outlines its role in the symbolism of spirituality." Even as technological advances in seventeenth-century Europe enabled astronomers to perceive the stellar nature of the Milky Way, the tendency was to regard this luminous apparition of the night sky as part of some grand theological construction. Even today, we cannot help but view the Milky Way as visceral proof of a vast cosmos that enmeshes us in its ethereal design.

Telescopic Revelations

To further understand the Milky Way, the astronomers of antiquity had to achieve several technological breakthroughs. These advances built upon one another until the limitations of our naked eyes were finally superceded by the insights gained from careful telescopic observations. This technological progression began with the early refinement of raw materials.

The manufacture of sand into glass takes us back once again to Near Eastern cultures, some 5,000 years ago. By cooking a mixture of silica and lime (obtained from sand) and soda (obtained from wood ash or sodium carbonate "natron" salts), the Egyptians and their Mesopotamian neighbors crafted decorative glass beads and figurines of multiple colors. Admixture of the soda was essential, as it enabled the sand to melt at feasibly low temperatures. According to the accounts of Pliny the Elder, the first realization of this technology occurred when sailors were cooking on a beach along the coast of modern Lebanon. The soda from blocks of natron which they used to support their cooking pots mixed with the sand on the beach to make globs of glass. In the next 3,000 years, the refinement of glassmaking spread throughout Egypt, Palestine, Syria, Mesopotamia, Greece, and Roman Italy.

The first optical magnifiers consisted of clear glass vessels filled with liquid. These were noted by the Roman playwright and philosopher Lucius Annaeus Seneca during the time of Christ. Seneca's notorious pupil Nero used an optically figured emerald as a monocle for viewing battles in the Coliseum. The crafting of optical-quality glass lenses is a relatively recent phenomenon, however, with the first magnifying lenses recorded by Roger Bacon in 1262 CE and the first eyeglasses appearing in Florence around 1280 CE.

The first lens-based telescopes are attributed to opticians in Holland, especially Hans Lippershey, who first wrote about the invention in 1608. These primitive instruments consisted of two lenses— the objective (receptive) lens being of convex (bulging) shape and the eyepiece lens being of concave (indented) shape. Together, the two lenses formed images of objects that revealed hitherto unseen details. These primitive refracting telescopes were sufficient to the task of magnifying views of distant ships and other objects of opportunity (or threat) beyond naked-eye reconnaissance.

The Italian physicist Galileo Galilei capitalized on the telescope's novel powers in his overtures to the powerful papacy. His hand-built spyglasses and their optical revelations were not unanimously embraced by the papal prelates, however. Galileo also exacerbated negative official reaction by promoting his telescopic views of celestial phenomena. From craters on the Moon, to dark storms on a rotating Sun, satellites revolving around Jupiter, and cyclic phases of the planet Venus, Galileo revealed a universe that was defiantly at odds with the standard doctrine of pristine planets in loyal orbits around the Earth. Instead, he found compelling evidence for the Sun ruling all planetary motion—a radical view that ultimately earned him house arrest.

When Galileo trained his small refracting telescope upon Orion, the Pleiades, and the Milky Way, he saw these celestial marvels resolved into myriad stars (see figure 2.3). As he wrote in his 1610 treatise *The Starry Message*:

> I have observed the nature and the material of the Milky Way. With the aid of the telescope this has been scrutinized so directly and with such ocular certainty that all the disputes which have vexed philosophers through so many ages have been resolved, and we are at last freed from wordy debates about it. The galaxy is, in fact, nothing but a congeries of innumerable stars grouped together in clusters. Upon whatever part of it the telescope is directed, a vast crowd of stars is immediately presented to view. Many of them are rather large and quite bright, while the number of smaller ones is quite beyond calculation.

Galileo's finding that the "smaller" stars were the most numerous foretells modern astronomy's continuing struggle to enumerate the census of stars throughout the Milky Way Galaxy. The history of this "star gauging" is one of fits and starts. Indeed, another 145 years would pass before any further progress would be made in discerning the stellar and nebular nature of the Milky Way.

The French Abbé Nicolas Louis de La Caille is regarded as the first to publish an extensive list of the "nebulae" that inhabit the night sky. Using a small refracting telescope of 1-inch lens diameter, he discovered and documented forty-two of these fuzzy apparitions. He made many of these discoveries during a voyage to the southern hemisphere, where he also collected data for an immense catalog of 9,776 stars. In his 1755 contribution "On the Nebulous

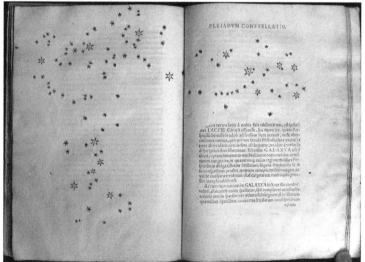

FIGURE 2.3. Galileo Galilei, surviving examples of the refracting telescopes that
he built, and his drawings of star fields in the Orion constellation (*bottom left*)
and the Pleiades star cluster (*bottom right*) c. 1610 CE. [(*Top left*) Painting by
Domenico Robusti (Tintoretto), c. 1605–07 CE. Image curated under Repro ID
BHC2699 © National Maritime Museum, Greenwich, London. (*Top right*)
Scanned image from *Studies in the History and Method of Science, Vol. II* by
C. Singer, Oxford, UK: Clarendon Press, 1921. Public domain. (*Bottom*) From
Siderius Nuncius (Starry Messenger) by G. Galilei, Venice, Italy, 1610 CE. Image
courtesy of History of Science Collections, University of Oklahoma Libraries]

Stars of the Southern Sky" to the *Memoirs of the Royal Academy of Paris,* La Caille proposed three classes of objects to explain his forty-two nebulae—(1) truly diffuse "spots" that resemble the bright nuclei of comets, (2) groupings of stars that appear nebulous to the unaided eye, and (3) stars with diffuse surroundings. La Caille had misgivings, however, wondering whether all three "species" could be explained by varied arrangements of stars, as Galileo had done in 1610. To do any better than La Caille, astronomers had to develop more powerful telescopes. The small refractors of the time were insufficient to the task, setting the stage for something new and better—the reflecting telescope.

Sir Isaac Newton is most commonly noted for being the father of gravitational physics. This prickly genius and his legendary insights from a falling apple have become icons of the Enlightenment Period (circa 1650–1800). We may never know whether or not Newton's gravitational theory was actually inspired by an apple falling on his head, but we do know that he was equally inspired by the quixotic nature of light. In his 1704 book *Optiks*, Newton proposed a particle theory for the propagation of light, portrayed the spectral character of light, and invented novel optical instruments. His 1668 design for a mirror-based telescope was one of the first to be successfully realized (see figure 2.4).

The eponymous Newtonian telescope consisted of a concave mirror of shiny metal that was placed at the bottom of a hollow cylindrical tube. The mirror gathered the incoming light and reflected it back up the tube to a pickoff mirror that redirected the light to an eyepiece on the side of the tube. The advantages of reflecting telescopes like those designed by Isaac Newton and his contemporaries James Gregory and Jacques (N.) Cassegrain were that one did not have to deal with large lenses and their inevitable imperfections. Only one surface had to be optically figured, while the material that made up the mirror's interior was inconsequential to the task at hand. Moreover, precisely ground mirrors did not suffer from the chromatic aberration (color-dependent blurring) that afflicted all lenses of the time.

Although Newton did not appear to do much astronomical observing, others in the eighteenth century made great use of his

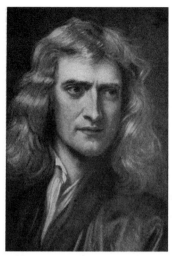

FIGURE 2.4. Sir Isaac Newton and his reflecting telescope c. 1668 CE. [Portrait originally painted by Sir Godfrey Kneller. Mezzotint by permission of Paul D. Stewart (Science Photo Library). Telescope curated at the Royal Society, London. Photo copyright Photos.com/Jupiterimages]

invention—making increasingly larger telescopes to perceive ever fainter sources of luminescence in the firmament. Several of these astronomers were motivated by the desire to discover—and name after themselves—the next bright comet. Toward these ends, the renowned comet hunter Charles Messier was continually frustrated by his observations of nebulous objects in the sky that masqueraded as comets. Using a reflecting telescope with a metal mirror of 7 ½-inch diameter and tube of 32-inch length, he set out in 1764 to find and catalog these "false comets." Together with his comet-hunting rival Pierre Méchain, Messier had compiled a list of 103 nebulous objects by 1784. We now realize that this assortment of sources includes clusters of stars and clouds of gas that reside in the Milky Way Galaxy—along with several spiral and elliptical galaxies that are situated well beyond our home galaxy. Many other non-stellar objects have been discovered in the intervening centuries. However, the so-called Messier Objects continue to command the attention of amateur and professional astronomers by virtue of their relative prominence through telescopes of modest aperture. Consequently, many of the beautiful objects that fill the pages of popular astronomy magazines and books are designated by their Messier or "M" numbers.

By the 1780s, enormous reflecting telescopes were being constructed and used in the service of mapping the heavens. The German-born Briton Sir William Herschel stands out as the most important astronomical instrumentalist and observer of these times. Having discovered the planet Uranus in 1781 with a reflecting telescope of 4-inch diameter aperture and 7-foot length, Herschel was supported by King George VI to devote himself to astronomical pursuits. He went on to build and deploy a 12-inch diameter, 20-foot-long reflector and a mammoth 48-inch diameter "40-foot" telescope. The "20-foot" telescope turned out to be the favored tool for surveying the Milky Way, as the larger telescope was far too difficult to operate. Together with his capable and devoted sister Caroline, William set out to "discern the construction of the heavens." Beginning with the 103 nebulae originally compiled by Messier and Méchain, the Herschels discovered and cataloged another 2,000 nebulae within the next seven years.

William vacillated throughout his long career on the true nature of the nebular objects that he cataloged with his sister. At first, he adhered to La Caille's view that he was beholding a mix of clustered stars and gaseous nebulae—too distant to be adequately discriminated. He even proposed that the true nebulae were tracing an evolutionary sequence of stars gravitationally condensing from clouds of gas. In his observations of the Orion Nebula, he correctly described it as "an unformed firey [*sic*] mist, the chaotic material of future suns." But he also described his so-called planetary nebulae as stars in the making. In his own words, "If this matter is self-luminous, it seems more fit to produce a star by its condensation than to have come from a star." We now know that these small nebulae are in fact the *end products* of stellar evolution, where dying stars are breathing their last gasps.

Perhaps of greater import to the study of the Milky Way itself, the Herschels' "star gaging" work led to the first quantitative mapping of the heavens. Sweeping the sky with his 20-foot telescope, William described his observations to Caroline, who carefully re-

FIGURE 2.5. (*Top*) Sir William Herschel and his sister Caroline. (*Middle*) The 20-foot telescope that they used to "discern the construction of the heavens." (*Bottom*) The Herschels' depiction of the Milky Way as viewed from beyond it (c. 1785 CE). The Sun is shown as a small dot in the middle of this flattened structure. In his later life, William Herschel recognized the limitations of his observations and imagined a Milky Way that extended to infinity. [(*Top left*) Portrait by Lemuel Francis Abbott (1785). Curated at the National Portrait Gallery, London, UK (NPG 98). (*Top right*) Portrait by Lisa Rosowsky (1987) based on various accounts of Caroline Herschel. Curated at the Harvard-Smithsonian Center for Astrophysics. Photo by W. H. Waller. (*Middle*) Woodcut image courtesy of the Royal Astronomical Society, Burlington House, Piccadilly, London, UK. (*Bottom*) From W. Herschel, *On the Construction of the Heavens*, in *The Scientific Papers of Sir William Herschel*, London: The Royal Society and the Royal Astronomical Society (1912)]

corded them along with the time of night and elevation of the telescope. From these records, they could reconstruct the number of stars in any direction of the sky. To derive a comprehensive map of the stellar distribution, William then made some broad assumptions. First, he let all stars have the same intrinsic luminosity. He did this in spite of having observed closely interacting double stars that were clearly disparate in brightness. He then let all stars be equally spaced from one another. From these two assumptions, his star counts provided a measure of how far the stellar distribution extended along any particular line of sight. The resulting map showed a clearly flattened structure, with a bifurcation that we may now recognize as the Great Rift in Aquila (see figure 2.5).

Later in his life, William Herschel cast doubt on his own nebular hypotheses and "the arrangement of the stars." Admitting the ocular limitations of his instruments, he made a final map of the Milky Way that showed a flattened system extending to infinity. In retrospect, we can now recognize that both of William Herschel's assumptions were in error. The intrinsic luminosities of stars vary by factors of more than a 100 million! Also, the spacing of stars is far from uniform. Nevertheless, we are beholden to William and Caroline Herschel for their pioneering use of telescopic observations to fathom our Galaxy as a distinct system of evolving stellar and nebular matter. Quoting the contemporary astronomer Charles Whitney, "Herschel had discovered a new continent in the sky; and he had discovered its enigmas but never resolved them: with courage and grace that are unique in the history of astronomy, he had dismantled his own constructions. . . . His work is a monument to the human spirit; he truly lives among the stars."

Back on Earth, William's only son John Herschel continued the family legacy by preparing his father's work for publication and by making his own astronomical observations. A versatile researcher and popularizer of science, John ventured to South Africa in 1833 for telescopic observations of the southern sky and for studies of the local flora and fauna. John's extensive observing led to the creation of the *New General Catalogue of Non-Stellar Objects*. The so-called *NGC* has since become a standard reference which today contains nearly 8,000 objects. Some of the nebulae that are mentioned in this book have both Messier and NGC listings. For example, the Orion Nebula is known as both Messier 42 and NGC 1976. Many other nebulae were missed by Messier and Méchain, however, and so are best known by their *NGC* numbers. The North America Nebula (NGC 7000) in Cygnus is a good example of a naked-eye nebula that did not make Messier's top 103 but found a place in Herschel's well-populated *NGC*.

Philosophical Perspectives

As the Herschels were plumbing the stellar and nebular Milky Way, philosophers of the eighteenth and early nineteenth centuries were creating their own cosmic constructs. We begin with Thomas Wright, an English theologian who advanced himself by teaching mathematics and physics to "noble ladies" of the time. In his 1750 book on *An Original Theory, Or New Hypothesis of the Universe*, Wright proposed that the Milky Way was part of a larger moral cosmos with the spiritual life force in its center. His model placed the Sun and Solar System inside a vast shell of similar stellar systems. From our vantage point, we would see innumerable stars along our part of the shell but much fewer stars perpendicular to it, thus explaining the concentration of starlight along the Milky Way (see figure 2.6). By having our local cosmos look slab-like, yet be part of a much larger spherical structure with supernatural power at its center, Wright was able to reconcile the physical with the transcendent. He went so far as to suggest that there may be an abundance of other spherical shell-worlds populating the greater universe. By 1770, however, he abandoned these notions, preferring to imagine the stars as distant volcanoes erupting from a spherical shell of rock that enfolds the Earth.

Contemporary with Wright, the Prussian Emmanuel Kant was making a name for himself through his tracts on perceiving reality. Kant's training in mathematics was far superior to that of Wright, and so when he read of the Milky Way being part of a flattened system, he conceived of a self-supporting disk of countless stars in shared circular rotation. Moreover, he could explain the "nebulous

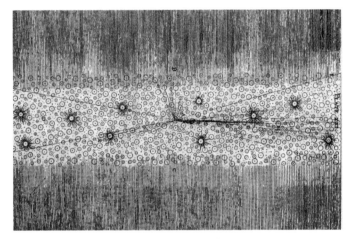

FIGURE 2.6. Thomas Wright's conception of the Milky Way as part of a spherical shell that surrounds some supernatural power. (*Top*) Wright imagined that our Earth-bound view of the Milky Way was caused by an excess of stars along the local slab-like portion of a much larger shell. (*Bottom*) The larger shell itself surrounded—at great distance—the divine center of "creation." [From T. Wright, *An Original Theory of the Universe*, London (1750)]

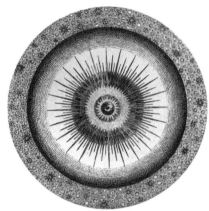

stars" that Messier and Méchain had found as other more distant "island universes" of similarly swirling stars. Although Kant's galactic views were not well known during his time, they were confirmed by the Herschels' observations, and have since developed into the modern paradigm of a vast universe populated by spiral and elliptical galaxies—each "island universe" in self-gravitating motion.

We end these philosophical reflections with Simon Laplace, a French mathematician who, more than anybody else, transformed our perceptions of the galaxian cosmos into a physically dynamic and evolving dynamo. In his six editions on *The System of the World* that were published between 1796 and 1835, Laplace considered the origin and development of the Solar System. He posited a "nebular hypothesis," whereby the Sun and planets condensed out of a swirling disk of nebular matter. Support for his theory came from the overwhelming propensity for the planets and their satellites to rotate in nearly the same plane and in the same direction. Laplace also had the mathematical wherewithal to show that Newtonian gravity would lead to these sorts of condensations. In brief, a rotating blob of gas will preferentially collapse along its axis of rotation, where the centrifugal forces are least able to support the nebula against collapse. Today, we refer to Laplace's nebular hypothesis, when we model the formation of other planetary systems in the Milky Way—and the birth of entire galaxies such as the Milky Way itself.

Stellar Feats

By the mid-1800s, astronomers equipped with telescopes, eyepieces, and notebooks had visually surveyed the entire sky. Extensive catalogs of stars were compiled from both the northern and southern hemispheres. The *Bonner Durchmusterung* (1860) (*Bonn Systematic Survey*) and its southern complement, the *Cordoba Durchmusterung* (1892), were the most comprehensive of these visual compilations. Observations of the northern hemisphere and a bit of the south were conducted at Bonn Observatory in Germany between 1852 and 1859, yielding a harvest of 320,000 stars down to tenth magnitude (where the brightest appearing stars are of first magnitude, the faintest stars visible to the naked eye are of sixth magnitude, and tenth-magnitude stars are another forty times fainter). In the southern hemisphere, the positions and visual magnitudes of 580,000 stars were recorded by astronomers at the National Observatory in Cordoba, Argentina. These monumental publications attest to the diligence of the visual observers who typically endured thousands of hours of grueling discomfort affixed to the eyepieces of their telescopes.

The pinnacle of visual observing was achieved on *individual* stars using the most precisely crafted telescopes of the time. The goal was to detect and measure a yearly shifting in a star's position—the reflexive response to our planet's orbit around the Sun. A handy analogy of this shifting or "parallax" is to place one of your thumbs in front of your face and alternately look at it with your left and right eye. Your thumb will appear to shift in position relative to the background scene. The more distant your thumb, the smaller the shift in its position. Astronomers were looking for similar shifts in stellar position as the Earth orbited from one side of the Sun to the opposite side. This quest to successfully measure a stellar parallax had been going on since Galileo's first telescopic reconnoitering of the heavens. A whole lot was at stake too, as the mere detection of a star's periodic parallactic shift would confirm that the Earth was in fact orbiting the Sun (and not the opposite), and that the star is indeed like the Sun but is at a vastly greater distance.

Success was finally achieved in 1838 by the German astronomer Friedrich Wilhelm Bessell, who made optimal use of an exquisitely precise refracting telescope. Crafted by the master optician Joseph von Fraunhofer, the specialized telescope had a split objective lens of 6-inch diameter which could be used to measure the tiniest of displacements between two stars (see figure 2.7). Bessell's targeted star was 61 Cygni, otherwise known as "the flying star" because of its large motion with respect to other stars. This motion was continuous rather than periodic and so was attributed to the star's trajectory in space or "proper motion" relative to the Sun. Stars with large proper motions were thought to be closer to us (which turns out to be true), and so 61 Cygni was regarded as a suitably nearby candidate for measurements of its periodic parallax.

After a year of scrutiny, Bessell obtained a parallactic shifting of a mere 0.31 arcsecond (where 1 arcsecond corresponds to 1/3600 of a degree). The distance of this star could then be calculated according to the defining formulation

$$\text{distance (parsecs)} = 1 / \text{parallax (arcseconds)},$$

where the distance in parsecs is defined to be that which produces a parallax of one arcsecond. For 61 Cygni, the distance amounted to 3.3 parsecs, or about 10 light-years (where 1 parsec equals 3.26 light-years). By meticulously measuring the position of just one

FIGURE 2.7. Friedrich Wilhelm Bessell and the specialized refracting telescope that Joseph von Fraunhofer crafted for him. Bessell used this exquisite instrument to measure the minuscule annual parallactic motion of the star 61 Cygni, thereby confirming the Earth's orbital motion around the Sun and so determining for the first time an accurate distance to anything beyond the Solar System. [(*Top*) Portrait from the Sternwarte, Universitat Bonn. Photograph courtesy of A. W. Hirshfeld. (*Bottom*) Engraving of telescope (the Koenigsberg heliometer) from Schweiger–Lerchenfeld Atlas der Himmelskunde (1898). Reference to *Parallax—The Race to Measure the Cosmos* by A. W. Hirshfeld, New York: W. H. Freeman/Owl Books (2001). Source: Widener Library, Harvard University, courtesy of A. W. Hirshfeld]

star, Bessell had finally confirmed the incredible vastness of the night sky—a firmament adorned with blazing suns at fantastic distances. In 1841, Bessell was awarded the Gold Medal of the Royal Astronomical Society. Sir John Herschel, then president of the RAS, gave the honorific speech—referring to the accomplishments of Bessell and his contemporaries as "the greatest and most glorious triumph which practical astronomy has ever witnessed." These words still ring true today.

Astrophysical Advances

By the mid-1800s, new sciences and technologies were emerging that would ultimately relegate the profession of visual "eyepiece" astronomy to the province of amateurs. Great strides in chemistry had led to the discovery of many new elements and molecular compounds. Moreover, the invention of the Bunsen burner and the spectroscope allowed scientists to ignite their chemical specimens and literally see what they were made of. By observing the light from the flames with a prism-based spectroscope, chemists could note the distinct patterns of colors emitted by sodium, magnesium, oxygen, and sundry other elemental substances. Spectroscopic examination of the Sun's incandescence revealed similar patterns. Instead of showing distinct colors in emission, however, the solar spectrum showed dark gaps in the broad "rainbow" of colors emitted by the Sun. First noted by the great optician Fraunhofer, these so-called absorption lines are now recognized as indicating the presence of hydrogen, oxygen, sodium, and iron in the Sun's fiery atmosphere (see figure 2.8).

By the 1860s, scientists were turning their spectroscopes to the stars. The British team of William Huggins and William Allen Miller led this new observational science, observing the spectra of such stellar luminaries as Rigel, Betelgeuse, and Vega, and finding that they differed dramatically in their respective spectral patterns (see figure 2.8). They also observed the Orion Nebula—discovering that its spectral lines were in emission rather than in absorption. Finally, the debate over the nature of the bright nebulae could be settled. Objects like the Orion Nebula were truly clouds of luminous gas rather than unresolved clusters of stars. Other nebulae could—in principle—reveal their inner constitutions through similar spectroscopic analysis. With these pioneering efforts, the modern science of observational astrophysics was born.

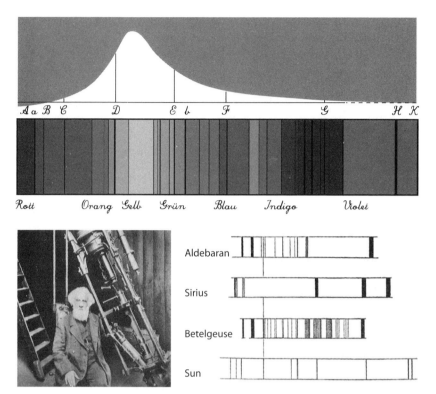

FIGURE 2.8. (*Top*) The spectrum of the Sun, as observed by Joseph von Fraunhofer, featured numerous dark "lines" at specific wavelengths. Similar absorption lines were found by William Huggins (*Bottom left*) and William Allen Miller in their pioneering spectroscopic studies of other stars, thus launching the field of stellar spectroscopy. (*Bottom right*) Stars spectroscopically observed by Huggins and Miller include, from top to bottom, Aldebaran, Sirius, Betelgeuse, and the Sun. The yellow sodium line at a wavelength of 589 nanometers appeared in three out of the four stellar spectra and so was used by Huggins and Miller to align the spectra. [(*Top*) From J. Fraunhofer, Denkschriften der Koniglichen Academie der Wissenschaften zu Munchen 1814–15, pp. 193–226. Redrawn for greater clarity. (*Bottom left*) From W. Huggins and M. Huggins, *The Scientific Papers of Sir William Huggins*, London: William Wesley and Son (1909). (*Bottom right*) Adapted from W. Huggins and W. A. Miller, "Note on the Lines in the Spectra of Some of the Fixed Stars," *Proceedings of the Royal Society* 12 (1863): 444]

Meanwhile, the chemical technology of photography was beginning to spread from its origins in France to the rest of the world. Before Mathew Brady recorded the American Civil War on wetted glass plates, the wealthy chemist and botanist John William Draper was making daguerrotypes of the Sun and Moon. This involved the cumbersome process of exposing a silvered copper plate to iodine vapors before inserting it in the camera, and then treating the plate to mercury vapors which would attach to the illuminated iodine. By 1880, John's gifted son Henry had produced photographic images of the Orion Nebula and other celestial phenomena using gelatin-treated glass plates that could withstand drying out during longer exposures and thus record much fainter features. He had also succeeded at obtaining photographic spectra of bright stars that clearly showed their characteristic absorption lines.

Following Henry Draper's premature death in 1882, his wife Anne was persuaded by William Pickering, Director of Harvard College Observatory, to set up a memorial fund for astronomical research. This fund supported a huge photographic survey of the sky, whereby the light from every star in the observed field of view was dispersed by a prism before being focused by an objective lens onto the photographic plate. The resulting 225,000 stellar spectra that were recorded on these plates laid the groundwork for a revolution in what we know about stars. As a living legacy, most stars brighter than ninth magnitude (sixteen times dimmer than the faintest naked-eye stars) are known by the "HD" num-

FIGURE 2.9. (*Top left*) William Pickering and his team of women at Harvard College Observatory who classified more than 200,000 stars according to their spectra as recorded on photographic plates. (*Bottom*) The spectroscopic sequence of stellar types that was developed by HCO's Annie Jump Cannon, Williamina Fleming, and Antonia de P. P. Maury, and that is still in use today. (*Top right*) Cecilia Payne-Gaposhkin, who correctly determined that the stellar spectroscopic sequence primarily tracked stellar surface temperature rather than chemical abundance. [(*Top left*) Harvard College Observatory Collection of Astronomical Photographs, published by B. L. Welther in "Pickering's Harem," *Isis* 73(1982): 94. (*Top right*) Courtesy of the Astronomical Society of the Pacific, with reference to C. *Payne-Gaposchkin: An Autobiography and Other Recollections*, by C. Payne-Gaposchkin, edited by K. Haramundanis Cambridge, UK:

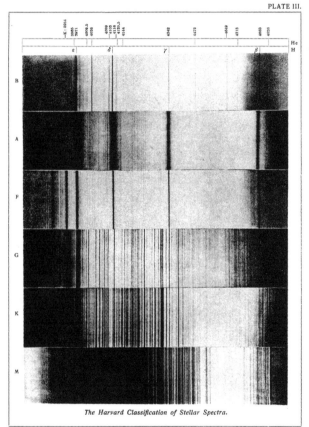

The Harvard Classification of Stellar Spectra.

Cambridge University Press (1984). (*Bottom*) Plate III from *Lectures on Stellar Statistics* by C.V.L. Charlier, Lund, Sweden: Scientia Publisher (1921). Scanned and published online under "The Project Gutenberg E-book of Lectures on Stellar Statistics Harvard Classification of Stellar Spectra"]

bers and spectral types that are listed in the voluminous *Henry Draper Catalogue.*

Much of Pickering's ambitious project was carried out by women—or "computers," as they were then called (see figure 2.9). Indeed, they achieved so much that it was impossible not to give them the credit they deserved. Today, we recognize Annie Jump Cannon, Williamina Fleming, and Antonia de P. P. Maury as the first true stellar diagnosticians. By classifying the many spectra according to the patterns of their absorption lines, and then re-sequencing these categories based on the stellar colors, these women paved the way for another woman to make sense of it all (see figure 2.9).

In her 1925 PhD thesis, Cecilia Payne took on the challenge armed with the latest theoretical understanding of atoms and their quantized workings.[1] Contrary to expectations, she found the differences in spectral types to trace the stars' surface temperatures rather than their chemical compositions. Stars are predominantly made of hydrogen (about 75% by mass), with some helium (about 23%) and a whiff of all the other elements that are evident in the spectra. It is the star's surface temperature that determines which element in the star's atmosphere is most receptive to absorbing the light from below. For a hot blue-white star such as Sirius, hydrogen is the primary absorber. This simplest of atoms will produce a unique pattern of gaps in the red, teal, and violet parts of the hot star's spectrum. A "warm" yellow star like the Sun will yield a spectrum rich with absorption lines of metals such as iron and magnesium. And a "cool" red star such as Betelgeuse will host absorption bands of molecules such as titanium oxide (see figure 2.10).

In less than forty years following the first spectroscopic studies of the brightest stars, astronomers could now understand the stellar horde in terms of physical quantities such as temperatures, luminosities, and chemical compositions. The "Rosetta Stone" of this stellar deciphering has become known as the Hertzsprung-Russell diagram (see figure 2.10). By relating a star's temperature to its overall luminosity (or equivalently its spectral type to its absolute visual magnitude), Ejnar Hertzsprung of Denmark and

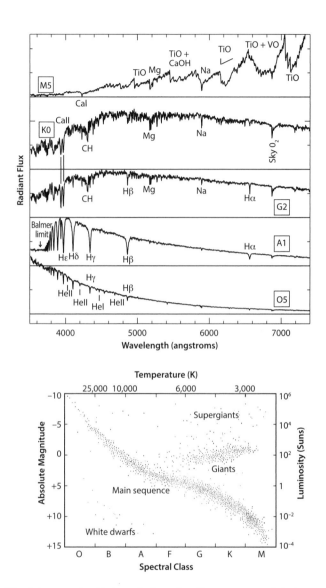

FIGURE 2.10. (*Top*) A more detailed rendering of stellar spectra and their corresponding spectral types. Here, a hot O-type stellar spectrum is shown at the bottom, a Sun-like G-type spectrum is in the middle, and a cool M-type spectrum is at the top. (*Bottom*) The eponymous Hertzsprung-Russell diagram which relates a star's spectral type (or equivalent surface temperature) to its absolute magnitude (or equivalent luminosity). At once, astronomers could see that stars come in three basic varieties: the so-called main sequence or dwarf stars, giants, and even more luminous supergiants. [(*Top*) Adapted from *Galaxies in the Universe* by L. S. Sparke and J. S. Gallagher, Cambridge: Cambridge University Press (2000), p. 205; courtesy of L. S. Sparke. (*Bottom*) Adapted by L. Slingluff from various sources]

Henry Norris Russell of the United States independently found distinct families among the stars.

Most stars fall along the so-called Main Sequence. These include the Sun, Vega, Sirius, and Rigel. Others occupy a distinctly more luminous tier known as the Red Giant branch. These include Capella, Arcturus, and Aldebaran. Still others were seen to blaze forth with luminosities exceeding that of the Sun by factors of 10–100 thousand. This Supergiant Branch of stars includes such powerhouses as Deneb, Betelgeuse, and Antares. In one diagram, astronomers could perceive essential differences among the stars—differences that have since been explained in terms of a star's mass and age. We will revisit these stellar revelations in chapters 5–8, when we consider the life cycles of the stars.

Nebular Conundrums

Progress on the nebulae came more slowly. Diffuse and dim, the so-called bright nebulae challenged both naked eye and photographic plate. Dispersal of their light into spectra further diluted their imprints and so compromised astronomers' ability to obtain any meaningful detections. The "spiral nebulae" were particularly vexing. Are they stellar "island universes," located far from our galactic environs, or are they gaseous whorls situated among the stars that delineate the Milky Way? This puzzle was posed by William Herschel and Emmanuel Kant in the late 1700s, yet remained intractable some 200 years later.

The "dark" nebulae posed their own conundrums. Noted since antiquity, these inky blotches and tendrils along the Milky Way confounded proper interpretation. Could they be vacant "windows" in the morass of stars, or are they distinct clouds of obscuring matter that are silhouetting the starlight from beyond? This question came to the fore with the deep photographs of Edward Emerson Barnard. Born into poverty in Nashville, Tennessee, Barnard first worked as a parlor photographer. His consuming interest in amateur astronomy prompted him to save up his meager savings and purchase a telescope of 5-inch diameter aperture. With this instrument, he proceeded to discover several comets. The monetary rewards from these popular discoveries enabled him to obtain an education at Vanderbilt University and employment at Lick Observatory in California beginning in 1887.

Observing with the 36-inch refractor at Lick and later the 46-inch at Yerkes Observatory in Wisconsin, Barnard discovered a record number of comets, the fifth moon of Jupiter, the star with the greatest known proper motion (now known as "Barnard's Star"), and a diffuse assemblage of stars which is cataloged as NGC 6822 but is now commonly called "Barnard's Galaxy." Despite these many accomplishments, it is his work with wide-field imaging that secured his great reputation among galactic astronomers. By improving upon the portrait cameras of his youth, Barnard obtained the deepest photographs yet of the Milky Way and other parts of the sky. Most of these photographs were obtained in 1905 from the clear dark skies of Mount Wilson in southern California. Barnard expended considerable effort in preparing the prints, which were ultimately published in 1927 as *An Atlas of Selected Regions in the Milky Way* (see figure 2.11).

In these prints, one can readily see the many dark nebulae that entranced and puzzled Barnard throughout his long career. He initially regarded these dark forms as holes and lanes in the starry firmament, but ultimately concluded that they were obscuring masses of unknown constitution. Having photographed the rich Sagittarius star fields during a partially cloudy night in 1913, he saw the resemblance between the effects of the foreground cumulus clouds and those being recorded on his deep photographic plates. In his own words, "I could not resist the impression that many of the black spots in the Milky Way are due to a cause similar to that of the small, black clouds mentioned above—that is, to more or less opaque masses between us and the Milky Way."

Most astronomers of Barnard's time essentially concurred with his basic conclusions regarding the dark regions in the Milky Way. More controversial was whether obscuring material permeated the

FIGURE 2.11. (*Top*) Edward Emerson Barnard at the business end of the 36-inch Lick refractor near San Jose, California. (*Bottom*) Photograph by Barnard of a selected region in Sagittarius. The image is rich with dark nebulae, many of which are known as Barnard objects. [(*Top*) Lick Observatory photograph. (*Bottom*) From *A Photographic Atlas of Selected Regions of the Milky Way* by Edward Emerson Barnard, ed. Edwin B. Frost and Mary R. Calvert. Washington, DC: Carnegie Institution of Washington (1927). Accessed from the Georgia Tech digital collections on E. E. Barnard at http://www.library.gatech.edu/search/digital_collections/barnard/index.html]

rest of interstellar space. Were Barnard's dark clouds distinctly unique entities, or was there a continuum of obscuration that affected everything we could see? This nagging question tainted every effort to fathom the true architecture of the Milky Way.

Kapteyn's Universe

The rise of photography in the nineteenth century had completely transformed the way astronomers carried out their research. By the late 1800s, visual atlases of the stars had been supplanted by photographic atlases. Of these, the *Cape Photographic Durchmusterung* led the way, with more than 450,000 stars of the southern sky recorded on photographic plates. Its success prompted French astronomers to replace the visual *Bonner Durchmusterung* with a new photographic survey of both the northern and southern skies which they would call the *Carte du Ciel*. This ambitious project involved eighteen observatories from around the world—each one charged with photographing a particular part of the sky in the same way. Seeing the need to carefully measure and analyze the resulting glut of photographic plates, the Dutch astronomer Jacobus Kapteyn convinced his home institution in Groningen to create an astronomical laboratory with no telescope—one that would be completely dedicated to the digestion of data that were acquired elsewhere.

Through careful analysis of many stellar positions over time, Kapteyn established a statistical relationship between a star's annual parallax and its proper motion. That meant he could use the mean proper motion of a group of stars as a proxy for the group's distance. Moreover, he found that some groups of stars were moving or "streaming" in special ways. The Galaxy was in motion, with some groups of stars passing clear through other vagabonding groups. These new findings inspired him to propose a new, even more ambitious *Plan of Selected Areas* that would marshal all that could be learned about the Milky Way's stellar constitution. By combining photographic photometry and spectroscopy, he would assemble complete dossiers on stellar positions, velocities in space, magnitudes, spectral types, and corresponding distances. The resulting census would advance our galactic perspective in ways that had not been achieved since the Herschels' initial "star gaging" a century earlier.

These lofty goals were eventually realized, though with seriously compromised results. The problem came with the age-old challenge of determining the distances to the stars. Kapteyn could apply the new astrophysics to his spectroscopic measurement of a star and so determine the star's spectral type. This stellar typecasting was often sufficient to the task of inferring the star's intrinsic (absolute) magnitude. A comparison of the star's absolute magnitude with its apparent magnitude would then yield the distance to the star. Although some distances would be in error, a large ensemble of distance determinations would yield reliable statistics on the true spatial distribution of stars. That is, if the presence of obscuring matter between the stars was ignored.

Kapteyn was fully aware of interstellar absorption as a potentially complicating agent. If present, it would yield fainter observed magnitudes. Comparison of a star's fainter apparent magnitude with its absolute magnitude would then result in an erroneously greater distance. Although Kapteyn vacillated over the extent and amount of interstellar absorption, he ended up deciding that it was inconsequential in regions away from the opaque clouds that Barnard had noted. By "grinding huge masses of facts into law," he obtained a universe of stars with a disk-like distribution that was centered on the Sun (see figure 2.12).

In retrospect, we can now recognize that Kapteyn's universe was beset by overestimated distances. Had Kapteyn properly accounted for the absorption by interstellar matter, the more distant stars in the Milky Way would have been much closer and more tightly packed. The resulting stellar distribution would then have been more uniform, with no significant falloff away from the Sun. Considering that Kapteyn's contemporaries were similarly confused by the uncertain interstellar situation in the Milky Way, we need not judge him too harshly. Yet, a simple naked-eye reconnaissance of the Milky Way's brightening near Sagittarius would have helped

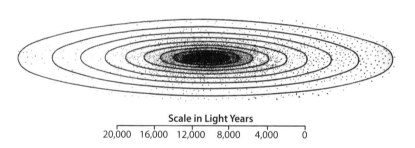

Scale in Light Years

20,000 16,000 12,000 8,000 4,000 0

FIGURE 2.12. Jacobus Kapteyn and the model of the universe that he derived based on having measured the distances to thousands of stars. The so-called Kapteyn universe was disk-like, with the Sun near its center. Many of the stellar distances were erroneously overestimated, however, producing the rapid (and spurious) fall-off in density away from the Sun. [(*Top*) Photograph courtesy of the Observatories of the Carnegie Institution of Washington. (*Bottom*) Adapted from P. J. van Rhijn, "J. C. Kapteyn Centennial," *Sky & Telescope* 10 (January 1951): 55. Based on findings of J. C. Kapteyn, "First Attempt at a Theory of the Arrangement and Motion of the Sidereal System," *Astrophysical Journal* 55 (1922): 302. © *Sky & Telescope*. Used with permission]

Kapteyn and his peers to appreciate that the Milky Way is asymmetric from our vantage point, with the direction toward Sagittarius appearing especially imposing. This naïve visual impression was ultimately confirmed by considering objects that do not confine themselves to our Galaxy's murky midplane.

Gauging the Galaxy

Henrietta Leavitt was one of William Pickering's able assistants at Harvard College Observatory (see figure 2.13). Educated at the women's college later known as Radcliffe, Leavitt began her career at the Observatory as an unpaid research assistant. She was soon charged with several key projects, including Harvard's involvement with Kapteyn's *Plan of Selected Areas*. Her work on variable stars, however, is what earned her place as one of the great astronomers of the twentieth century. By carefully examining photographic plates of the Large and Small Magellanic Clouds that had been obtained between 1904 and 1908, she identified more than 1,700 variable stars. The brightest of these were known as Cepheid variables—named after the prototype yellow supergiant star Delta Cephei.

FIGURE 2.13. (*Top*) Henrietta Leavitt and the Small Magellanic Cloud, where she found 1,777 Cepheid variable stars and measured their periods of varying brightness. (*Middle*) The characteristic "sawtooth" light curve of the prototype Cepheid variable star, Delta Cephei. (*Bottom*) The key relation that Leavitt discovered between a Cepheid's period of variability and its luminosity. Once calibrated with nearby Cepheid variables of known distance, this relation could then be used to gauge the distances of more remote Cepheids in the Milky Way and other galaxies. [(*Top left*) Courtesy the American Association of Variable Star Observers (AAVSO). Accessed from Wikipedia Commons. (*Top right*) Davide De Martin and the Digitized Sky Survey 2, European Space Agency and Hubble Space Telescope, NASA. (*Middle*) Adapted from *Galaxies and the Cosmic Frontier* by W. H. Waller and P. W. Hodge, Cambridge, MA: Harvard University Press (2003), with reference to *Galaxies* by Harlow Shapley, Cambridge, MA: Harvard University Press (1943, 1961, 1972). (*Bottom*) Adapted from *Galaxies and the Cosmic Frontier* by W. H. Waller and P. W. Hodge, Cambridge, MA: Harvard University Press (2003) with reference to p. 61; from H. S. Leavitt, *Harvard College Observatory Circular* no. 173 (1912), p. 1]

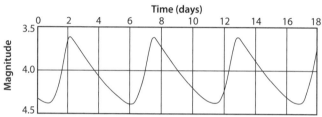

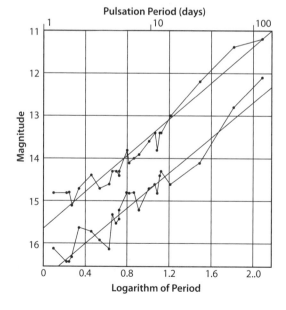

Leavitt understood that all of the Cepheid variables in the Magellanic Clouds were at essentially the same distance. Therefore, any differences in mean brightness among them could be attributed to actual differences in intrinsic luminosity. By plotting each star's mean brightness against its period of variation, Leavitt came upon a seminal relation. The more luminous Cepheids were found to vary more slowly than their dimmer cousins (see figure 2.13). This relation could be turned around, so that one need only track a remote Cepheid's period of variability to infer its intrinsic luminosity. And by relating the star's luminosity to its apparent brightness, one could reckon its distance. With the Cepheids, Leavitt had discovered powerful "standard candles" that could be used to fathom distances a million times greater than that possible with the parallax method, and a thousand times greater than the spectroscopic distances measured by Kapteyn.

The first objects to get their distances estimated in this way were the Magellanic Clouds themselves. Ejnar Hertzsprung calibrated the period-luminosity relation using nearby Cepheids, whose distances could be estimated by independent means. He then applied this calibration to the Magellanic Clouds, obtaining a distance of about 30 thousand light-years. This initial estimate was erroneously published as 3,000 light-years, making the Clouds a part of the Milky Way rather than the separate galaxies which they are. Even so, Hertzsprung's mis-reported estimate opened the way to further improvements and consequential applications.

Harlow Shapley, a Missourian with an early background in journalism, happened upon his illustrious career in astronomy by serendipity. Matriculating at the University of Missouri in 1907, Shapley expected to pursue his interests in journalism. However, he found that there were no journalism courses then available at the university, and so he looked for an alternative in the course catalog. He found Astronomy under the A's immediately after Archaeology—which he could not pronounce—and pursued the more pronounceable field as his major study. From Missouri, Shapley went on to study at Princeton under Henry Norris Russell, where he made thousands of observations of variable stars and computed the orbits of ninety eclipsing binary star systems for his doctoral thesis. These latter measurements were critical to learning what little was then known about the sizes and masses of stars. By closely monitoring the eclipses between closely dancing star couples, he found that red giants were, in his own words, "enormous gas bags," hundreds of times larger than the Sun.

Shapley's work on eclipsing binaries led him to study Cepheid variables, as they were then thought to be a type of eclipsing binary. Their unique pattern of variation was difficult to reconcile with the binary model, however, and so he and Russell began to consider Cepheids as pulsating variable stars rather than binary systems. By 1912, Henrietta Leavitt had established the Cepheid's period-luminosity relation. Again, models of eclipsing binaries failed to predict this relation, whereas models based on pulsating stars seemed to do much better. In this model, a Cepheid's luminosity would systematically vary, as its size periodically bloated and shrank. The success of this model emboldened Shapley and his colleagues to use Cepheids as "standard candles" for estimating distances to globular clusters.

Shapley's pursuit of globular clusters followed the time-honored principle of uniformity. In his observations and subsequent measurements, he first assumed and then sought to confirm that the Cepheids in globular clusters, the Milky Way's disk, and the Magellanic Clouds all behaved the same. In so doing, he determined distances to about sixty-nine globular clusters. These appeared to cluster in the direction of Sagittarius (see figure 2.14). From his distance measurements, Shapley showed that the spatial distribution of globular clusters was in fact centered about a point in the direction of Sagittarius—about 60,000 light-years away from the Sun. The offset location of this distribution prompted Shapley to argue for an enormous "Metagalaxy" that was 300,000 light-years in extent, with the Sun no longer residing in the center. Shapley's displacement of the Sun away from the center of his metagalactic system was revolutionary—recalling the last major shift from a geocentric to a heliocentric cosmology that Copernicus proposed some 400 years earlier.

The controversy over the Milky Way's "sidereal structure" culminated with a well-publicized debate between Shapley and Heber Curtis—an astronomer from Lick Observatory who had extensively studied the vexing "spiral nebulae." Sponsored by the Na-

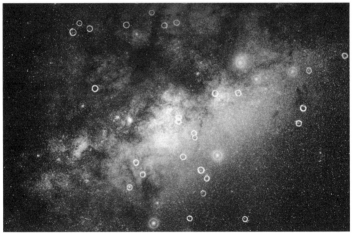

FIGURE 2.14. (*Top*) Harlow Shapley at his gigantic revolving desk. (*Bottom*) Shapley's identification of globular clusters in the direction of Sagittarius. Shapley's model of the Metagalaxy was based on measurements of Cepheid variable stars in about 69 globular clusters. The centroid of his globular cluster distribution was offset from the Sun by about 60,000 light-years in the direction of Sagittarius. [(*Top*) From *Ad Astra Per Aspera, Through Rugged Ways to the Stars* by Harlow Shapley, New York: Charles Scribner's Sons (1969); image courtesy AIP Emilio Segre Visual Archives, Shapley Collection. (*Bottom*) Harvard College Observatory; published in *Star Clusters* by H. Shapley, New York: McGraw-Hill (1930), and *Through Rugged Ways to the Stars* by H. Shapley, New York: Charles Scribner's Sons (1969)]

tional Academy of Sciences, the so-called Great Debate was held in 1920 in Washington, DC.

Shapley argued for his offset model, where he promoted the distribution of globular clusters as a superior tracer of the Milky Way's overall shape. He also considered the spiral nebulae to be nearby parts of his expansive metagalactic system. In these arguments, he was swayed by the observation of an unusually bright nova in the Andromeda Nebula. If the Andromeda Nebula and other spiral nebulae were separate and distant "island universes," then the bright novae that they occasionally displayed would be impossibly luminous. Therefore, they are most likely nearby parts of the metagalaxy. Shapley was also influenced by rotational motions in the spiral nebulae that had been measured and documented by his Mt. Wilson colleague and friend, Adriaan van Maanan. By carefully examining photographic plates of spiral nebulae that had been taken over several years, van Maanan had found systematic shifts in the positions of particular small features. These amounted to a few one-hundredths of an arcsecond per year, in a rotational pattern that could only be sustained if the nebulae were relatively nearby. Were these nebulae isolated "island universes" at distances exceeding a million light-years, the tiny angular displacements would translate to impossible rotational velocities that exceeded the speed of light. Therefore, Shapley concluded that the spiral nebulae and everything else had to be relatively nearby denizens of his metagalaxy.

Curtis concentrated on the spiral nebulae, arguing that the occasional nova outbursts in them were intrinsically much brighter than the more frequent novae found within the Milky Way. Being ultra-luminous, these exceptional novae were being observed in spiral nebulae that were at great distances from us. He also doubted Shapley's comparison of stellar "standard candles" among the Milky Way, globular clusters, and Magellanic Clouds—suspecting that the stars were in fact very different in intrinsic luminosity, and hence were at distances that would make the Milky Way smaller and the spiral nebulae of equivalent size but much farther away. Curtis was unconvinced by the rotational motions observed by van Maanan, having heard of contradictory results by other scientists. He was more impressed by the distribution of the spiral nebulae which completely avoided the murky Milky Way and the spectra of these objects which showed absorption lines typical of stars. To Curtis, these qualities indicated autonomous stellar systems far from the Milky Way.

In retrospect, we see both truth and error among these two debaters. Shapley was right about the globular clusters tracing the overall shape of the Milky Way—and the use of Cepheid variables in fathoming the distances to these clusters. However, his "standard candles" were not nearly as standard as he had assumed. The so-called Cepheids in the globular clusters turned out to be ten times less luminous than those in the Milky Way's disk and in the Magellanic Clouds. Accounting for this discrepancy places the globular clusters closer and hence makes his metagalaxy smaller by a factor of 3 or so. Van Maanan's rotational motions turned out to be spurious consequences of very difficult measurements. We can appreciate today that the spiral nebulae were much too far away for any rotational motions to be discerned. And the bright novae in the spiral nebulae were in fact thousands of times more luminous than the novae observed in the Milky Way. These "supernovae" were erupting from stellar systems millions of light-years more distant than the Milky Way.

A Galaxy among Galaxies

Indisputable evidence for these conclusions was found by Edwin Hubble in 1924. A graduate of the University of Chicago and Oxford University, Hubble was a physically vigorous and somewhat pompous personality with strong ambitions. Although he was schooled in law according to his father's wishes, he was most earnest about learning astronomy. Following his father's death, Hubble indulged his abiding passion at the University of Chicago, where he began photographing faint nebulae under the tutelage of Edward Emerson Barnard—by then one of America's most renowned astronomers. Hubble's photographs led him to classify the various nebulae according to their shapes. We inherit the elliptical, spiral, and irregular classifications of galaxies from these pioneering efforts (see figure 2.15). Later, at Mt. Wilson Observatory, he obtained spectra of these nebulae, showing that most of them were moving away from the Milky Way. Like Curtis and others before him, Hubble interpreted these motions as arising from "stellar systems at distances to be measured often in millions of light-years."

Hubble's spectroscopic studies also allowed him to discriminate between the truly nebulous objects in the Milky Way and the stellar assemblages which he suspected were much farther away. His insights on the gaseous nebulae were on the mark—noting that "the nebulosity has no intrinsic luminosity, . . . but either is excited to emission by light from a star of earlier [hotter] type or merely reflects light from a star of later [cooler] type." For the nebulae with stellar spectra, the big breakthrough came when Hubble resolved a variable star in the Andromeda Nebula. At first, he thought it was the latest of several novae which he had discovered in this system. However, further inspection of photographic plates that had been taken of the same region revealed to him that the source had dimmed and brightened again—contrary to any known nova eruption. Upon analyzing several more nights of observation, Hubble had his discovery. The source was a Cepheid variable star, whose apparent brightness placed it and its hosting nebula some 825,000 light-years away. Shortly thereafter, Hubble wrote a letter to Harlow Shapley at Harvard College Observatory, sharing the epochal news. Upon reading the note, Shapley was heard to say, "Here is the letter that destroyed my universe."[2]

Hubble continued to look for Cepheid variables in Andromeda and other large nebulae. He found thirty-five Cepheids in Messier 33 and eleven Cepheids in NGC 6822—the diffuse assemblage first noted by E. E. Barnard. The corresponding distances and sizes of these systems showed them to be autonomous Milky Ways— what Shapley later called "galaxies." Today, we continue to regard the Milky Way as a robust but representative member of a vast galaxian universe.

Within the Milky Way itself, studies of open star clusters by Robert Trumpler at Lick Observatory finally confirmed the ubiquity of interstellar dust (see figure 2.16). Its presence dimmed the light from these clusters so that they appeared more remote than

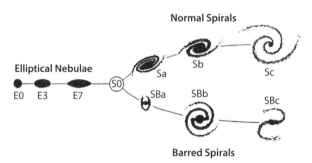

FIGURE 2.15. (*Top*) Edwin Hubble at the 48-inch Schmidt camera on Palomar Mountain. (*Bottom*) Hubble's 1924 classification scheme for the "nebulae" that he had photographed at Yerkes Observatory and Mount Wilson Observatory. Within the next five years, Hubble determined that these systems were indeed autonomous galaxies, millions of light-years more distant than anything in the Milky Way. [(*Top*) Caltech Archives Photo ID 10.12–17 by permission. (*Bottom*) Adapted from *The Realm of the Nebulae* by E. Hubble, New Haven, CT: Yale University Press (1936), p. 45]

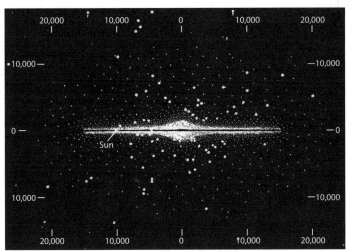

FIGURE 2.16. (*Top left*) The Swiss-American astronomer Robert J. Trumpler discovered in 1930 that stellar distances in the galactic disk had been overestimated due to the obscuring effects of interstellar dust. (*Top right*) In 1936, the Canadian astronomer John S. Plaskett synthesized the findings of Shapley, Trumpler, and others—imagining a flattened centrally concentrated disk of stars, gas and dust that is surrounded by a halo of globular clusters. (*Bottom*) Plaskett's edge-on rendering of the "galactic system," where the spatial dimensions are noted in parsecs. With some exceptions, this model of the Milky Way's stellar and dust components is consistent with what we know today. [(*Top left*) Courtesy of the Mary Lea Shane Archives of Lick Observatory, University of California, Santa Cruz. (*Top right*) Courtesy of the Dominion Astrophysical Observatory, Hertzberg Institute of Astrophysics, National Research Council of Canada. (*Bottom*) Adapted from Plate XIV in "The Dimensions and Structure of the Galactic System" by J. S. Plaskett in the *Journal of the Royal Astronomical Society of Canada* 30, no. 5 (1936): 153–164]

the distances that he reckoned based on their respective angular sizes. Trumpler estimated a mean dimming of about a 0.7 magnitude per kiloparsec of distance through interstellar space. This was enough to transform Kapteyn's Sun-centered universe into a much smaller version of Shapley's metagalaxy, with the Sun residing some 10,000 light-years off-center.

By the 1930s, the basic stellar specifications of our home galaxy had been pretty much established. The Canadian astronomer John Plaskett summarized the new consensus in a schematic drawing that he published in 1936 (see figure 2.16). In this edge-on view, we can clearly identify the Sun amid a dusty disk of stars whose dense center is 10,000 parsecs (33,000 light-years) away. The disk and central bulge are surrounded by dozens of globular clusters which collectively trace the so-called halo. Today, astronomers may argue over the absolute dimensions of this picture. For example, the distance between the Sun and galactic center is currently reckoned at about 8,500 parsecs (27,700 light-years). However, most astronomers would feel fairly comfortable with the overall composition of Plaskett's rendering.

Since the 1940s, dramatic enhancements in our understanding of the Milky Way have ensued. From giant clouds of molecular gas to odd goings-on in the galactic center, most of these new insights have benefited from observations at nonvisible wavelengths. In the next chapter, we will consider the stunning vistas and wondrous realms that these multi-wavelength observations have revealed.

CHAPTER 3

· · ● ● ● · ·

PANCHROMATIC VISTAS

To know ultraviolet, infrared, and X-rays
Beauty to find in so many ways.
—In "The Word," by Graeme Edge and the
Moody Blues (1968)

IT IS STRANGELY UNSETTLING to peer through a telescope at faint patches of starlight and nebulosity. Despite the modern telescope's extraordinary powers, the stiletto stars defy magnification. Instead, the telescopic viewer is confronted with starscapes of numerous shimmering points—each point representing a completely inscrutable solar system. The nebular smudges do manage to resolve into diaphanous forms—as colorless and ethereal as phantoms on a midnight romp. Distant, myriad, and uncaring, these uncanny sights elude human negotiation. No wonder so many people choose to keep their porch lights on.

Yet this same aloofness is what draws astronomers to poke and probe with ever greater intention. What structures can be perceived in these fantastical realms? If we could amplify our senses a trillion-fold, what would be the true mix of colors? And what could the observed structures, colors, and motions tell us about life and death among the stars?

To answer these questions, astronomers have enlisted—and advanced—the latest technologies in an unending quest to see the previously unseen. Beginning with telescopic views sketched on notepads, astronomers have pushed photographic technologies to their practical limits, progressed to recording individual photons of

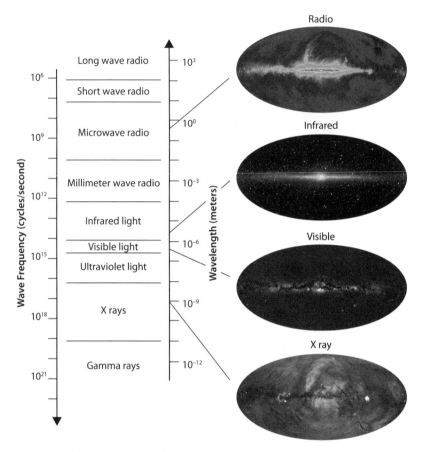

FIGURE 3.1. The electromagnetic spectrum with all-sky views at representative wavelengths. [All-sky images from the Multiwavelength Milky Way website hosted by NASA's Goddard Space Flight Center at http://mwmw.gsfc.nasa.gov/]

light with solid-state electronic array detectors, and extended their chromatic sensitivities so that the entire electromagnetic spectrum is now open to scrutiny. In studies of the Milky Way, this latter panchromatic facility has made all the difference (see figure 3.1).

The Radio Sky

The radio end of the electromagnetic spectrum spans wavelengths of centimeters to tens of meters and corresponding frequencies of gigahertz to kilohertz. It was the first nonvisible form of light to be mapped across the sky. In 1933, radio engineer Karl Jansky crafted a radio antenna made of wire and wood for the purpose of investigating the feasibility of radio-based telephone communications across the Atlantic Ocean (see figure 3.2). When he deployed his rotating antenna at Bell Labs in Holmdel, New Jersey, he found that the radio noise from the sky rose and fell every twenty-four hours. This cyclic behavior persisted for several months—long enough for Jansky to rule out the Sun as the primary source. Upon comparison with star charts, he determined that the most intense times occurred when the Milky Way was highest in the sky. Few scientists took notice of Jansky's discovery, however, and the Great Depression quelled further progress. That is until 1937, when Grote Reber of Wheaton, Illinois, built a fully steerable 30-meter diameter radio dish antenna in his own backyard. By the early 1940s, Reber's radio observations had revealed the Milky Way itself, with the region toward Sagittarius shining brightest (see figure 3.2).

World War II quickly propelled radio and radar technologies to the forefront, inadvertently putting radio astronomy on the fast track. By the 1950s, astronomers were working with 100-foot diameter radio dishes that could capture faint signals from the Milky Way and beyond. Today, these large radio telescopes have produced large-scale maps of the radio "noise" at angular resolutions of about 1 degree. Although this resolution is much poorer than that of the human eye, it is sufficient to show that the Milky Way is alive with matter that has been excited to glow at radio wavelengths (see plate 6).

Radio Continuum Emission

The cosmic radio emission first tracked by Karl Jansky, Grote Reber, and others has a smooth spectrum of intensities—what is known as continuum emission. Most of this emission comes from free electrons in the interstellar medium. Some of the negatively charged electrons are zipping past positively charged ions of hydrogen (consisting of a sole proton) and helium (with 2 protons and 2 neutrons)—the two most abundant elements in the cosmos. As the electrons career by these charged nuclei, they give up energy in the form of *Bremsstrahlung radiation*. This form of light is

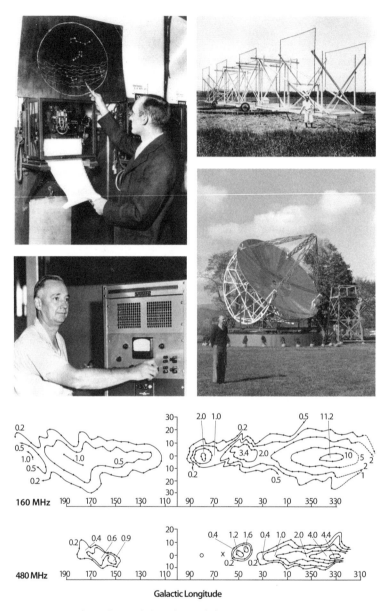

FIGURE 3.2. (*Top*) Karl Jansky with his chart of the radio sky and the radio telescope he built. (*Middle*) Grote Reber at his controls and his steerable radio telescope. (*Bottom*) Reber's mapping of the Milky Way at two radio frequencies (re-drawn for greater clarity). Galactic longitude runs along the *x*-axis and galactic latitude along the *y*-axis. Note that the brightest region was observed at a galactic longitude of 330°. In 1958, galactic longitudes were shifted 30 degrees so that radio mappings like these would peak at 0°/360°, thus defining the galactic center. [Images courtesy of the National Radio Astronomy Observatory (NRAO), Associated Universities Incorporated (AUI), National Science Foundation (NSF). They are archived at http://images.nrao.edu/Historical]

readily detected at radio wavelengths. It is present wherever electrons have been liberated from their host atoms, a situation which usually involves temperatures of several thousand degrees Kelvin. The ionized nebulae that surround caches of newborn hot stars are strong Bremsstrahlung emitters, as they have the right sorts of temperatures and are of sufficiently high densities to loudly pronounce their presence.

Other free electrons are traveling at incredible speeds close to that of light. These "relativistic" electrons can be found spiraling around magnetic field lines in the intense "magnetospheres" of rapidly spinning neutron stars, in the torrid remnants of supernova explosions, and throughout the diffuse interstellar medium. The situation is analogous to that of particle accelerators, where electrons and other charged particles are steered by magnetic fields into high-speed circular trajectories. The electrons in both situations lose energy—in what is known as *synchrotron radiation*. We can observe this radiation beaming from pulsars, supernova remnants, and wherever else relativistic electrons and magnetic fields are intertwining.

Radio Line Emission from Atomic Hydrogen

Another form of radio emission has become especially important to astronomers, as it traces the single most abundant gas in the cosmos. Atomic hydrogen consists of a single positively charged proton with a negatively charged electron swarming around it. The electron's orbit is quantized, yielding discrete energy levels that the electron can occupy according to its degree of excitation.

The electron also has a quantum property akin to spin. When the electron is in its favored state of lowest energy, its spin is opposite to that of the hosting proton. However, it doesn't take much to excite the electron to spin in parallel with the proton. When the electron eventually flips back to its lower energy spin state, it emits a photon of very low energy. The corresponding radio emission has an oscillatory wavelength of 21 centimeters and frequency of 1420.4 million cycles per second (mega-Hertz, or MHz). In 1944, Dutch astronomer Hendrik van de Hulst predicted that this special emission would pervade the Galaxy. Six years later, it was discovered by Harold Ewen—a graduate student working in the labora-tory of Harvard physicist Edward Purcell. Since then, giant arrays of radio dishes have been built with the express purpose of map-ping the 21-centimeter emission line of atomic hydrogen with ever greater sensitivity and acuity.

As shown in plate 6, the 21-centimeter radio emission from atomic hydrogen is brightest near the midplane of the Milky Way. Because it traces the most abundant element in the cosmos, the 21-centimeter emission provides an excellent tracer of the gas that dwells in the disk of our Galaxy. Another virtue of this emission is that it is unimpeded by the presence of dust. The long-wavelength radiation handily wriggles through any intervening clouds of dust—allowing us to perceive gaseous regions clear to the far side of the Galaxy. Perhaps most important, the spectral-line character of the emission is sensitive to relative motions along the line of sight. Ac-cording to the Doppler effect, the radio waves from receding gas clouds will be stretched to wavelengths greater than 21 centimeters. This increase in wavelength can be measured to great accuracy with radio receivers, enabling astronomers to clock the speeds of the re-ceding clouds. The same holds true for approaching gas clouds. Here, the wavelength of received radiation is compressed to values smaller than the fiducial 21 centimeters, the degree of compression being in direct proportion to the velocity of approach.

Radio Reconnoitering of Dark Matter

By measuring the wavelengths of peak emission from atomic hy-drogen all around the Galaxy and analyzing these measurements according to the Doppler effect, astronomers have learned how the hydrogen gas is moving about the Milky Way. They found that the gas clouds are in nearly circular orbits around the galactic center. Much to their surprise, however, the orbital velocities do not de-crease at greater radii from the galactic center. Instead, the veloci-ties remain essentially constant out to radii of more than twice the Sun's orbital radius of 27,000 light-years (see figure 3.3).

Contrast this situation with that of the Solar System, where the planets orbit at velocities that decrease with the square root of

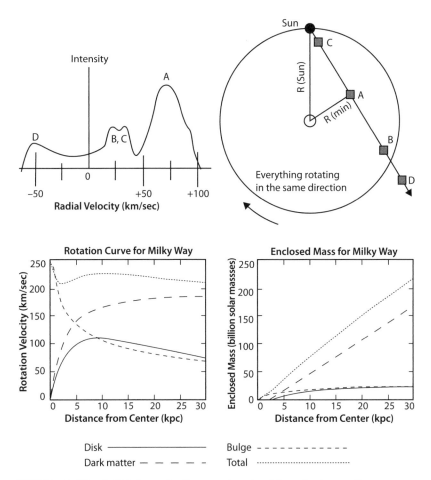

FIGURE 3.3. (*Top left*) Schematic diagram showing 21-cm hydrogen-line spectrum from a line of sight passing through the inner galaxy. (*Top right*) This sight line intersects multiple sources of emitting hydrogen. The highest velocity source of emission (A) comes from gas that directly traces the rotational speed at the source's galactocentric radius [R(min)]. (*Bottom left*) By sampling multiple lines of sight through the inner galaxy, and combining the kinematic results with observed motions in the outer galaxy, astronomers have derived a rotation curve for the entire galaxy that is remarkably constant with radius (top curve). Gravity from the observed bulge and disk components is insufficient to bind these motions. (*Bottom right*) Instead, the rotational velocities indicate an ever increasing amount of gravitating mass as a function of galactocentric radius. This mass is thought to be in the form of a Dark Matter halo that vastly outweighs all other visible matter. [Adapted from the online astronomy textbook by Nick Strobel archived at http://astronomynotes.com]

their respective distances from the Sun. In the case of the Solar System, the planets are responding to the presence of a single dominant mass—the Sun. The orbital velocity of each planet can be readily understood as being in equilibrium with the gravitational binding between the planet and the Sun. Now reconsider the swirling gas clouds in the Milky Way. They are traveling much faster than would be expected from a single dominating central mass. In fact, they are zooming around the Galaxy at velocities that defy gravitational binding by *any* known form of matter in the Milky Way. There must be something more keeping these clouds from flying off into intergalactic space. That ineffable something has been dubbed "*dark matter*," as it must be ponderous and yet unseen—at any wavelength.

As shown in figure 3.3, the velocities of hydrogen in the Milky Way indicate an ever-increasing amount of enclosed mass as a function of galactic radius. The observed disk and bulge components max out at about 25 billion solar masses. We are left to infer the presence of a dark matter halo that keeps increasing with radius—yielding an estimated total of more than 300 billion solar masses.

The nature of the dark matter remains one of the greatest enigmas in modern Astronomy. Evidence for it can be found in the excessively rapid stellar and gaseous motions within the Milky Way and within other galaxies, in the zippy relative motions between galaxies in galaxy clusters, and in the strength of gravitational lensing produced by galaxy clusters on more distant background galaxies. In the latter instance, images of the background galaxies are distorted by the intense gravity inherent to the foreground clusters. All of these phenomena indicate amounts of dark matter that dwarf all substances that can be directly detected. In the Milky Way, modeling of the orbital velocities and estimations of the stellar and nebular masses indicate that more than 80 percent of the Galaxy's mass is in some dark form. What that form might be remains completely unknown.[1]

Radio Searches for Extraterrestrial Intelligence

Besides revealing what we cannot see—and what we cannot even detect in a direct manner—the radio portion of the electromag-

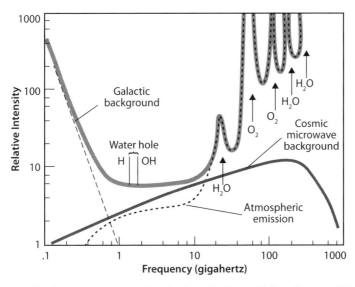

FIGURE 3.4. Radio spectrum showing the Cosmic Water Hole, where artificial signals from extraterrestrial civilizations would be most detectable. [Adapted from *Astronomy—A Beginner's Guide to the Universe*, 6th edition by E. Chaisson and S. McMillan, San Francisco: Pearson Education (2010)]

netic spectrum provides a relatively "quiet" zone, where emission from the Earth's atmosphere and natural sources in the Milky Way are at a minimum (see figure 3.4). The key exceptions are the emission lines of atomic hydrogen at 21-centimeter wavelength (1420 MHz frequency) and of the hydroxyl radical (OH) at 18-centimeter wavelength (1660 MHz frequency). If you put those two substances together, you get water—the veritable wellspring of life. For this reason, the first major searches for extraterrestrial intelligence (SETI) targeted this particular region of the radio spectrum. It was quiet enough to look for artificial signals from somewhere else in the Galaxy, and it would be understood by other water-based life as being special. Bernard Oliver, the first chief of NASA's SETI program, dubbed this spectral region the "Cosmic Water Hole." "Where shall we meet our neighbors?" he asked. "At the water-hole, where species have always gathered." Today, the radio portion of the electromagnetic spectrum continues to be the dominant hunting ground for possible transmissions from technologically communicative extraterrestrial lifeforms.

The Microwave Sky

The shorter-wavelength and higher-frequency portion of the radio spectrum requires precisely figured radio dishes in order to properly focus the radiation. The surfaces of these dishes are akin to smooth cooking pans but with diameters that span 10–100 meters. The detectors are also a lot more sophisticated than the receivers that can be used at longer radio wavelengths. Instead of coaxial cables transmitting the signals, tubular "waveguides" and wire-mesh "optics" have to be configured to direct the radio waves to their intended destination. For these reasons, millimeter-wavelength (or microwave) astronomy did not make much headway until the late 1970s. It then took off, with new discoveries coming in rapid succession. Most of these discoveries were of circumstellar or interstellar molecules. By the 1980s, hundreds of different molecules had been discovered through their emission at millimeter and centimeter wavelengths. Most of this emission occurs when the molecule changes its rate of rotation from one quantized energy level to another lower level. Table 3.1 lists some of the more important molecules found in our Galaxy.

The work horse of molecular Astronomy has been the carbon monoxide (CO) molecule. Unlike the abundant but faint diatomic molecule of hydrogen (H_2), carbon monoxide is readily excited to glow—even at very low temperatures. The molecule's brightest emission occurs during the transition from its first excited rotational level to the ground level. The resulting radiation can be observed at a wavelength of 2.6 millimeters and corresponding frequency of 113 billion cycles per second (Giga-Hertz or GHz). As shown in plate 6, the CO emission hews closely to the galactic midplane. The few notable exceptions arise from nearby molecular clouds whose modest displacements from the midplane are magnified into large apparent deviations due to their relative proximity.

Thanks to the generously luminous CO molecule, astronomers can map the distribution of molecular clouds throughout the Galaxy. These clouds are huge and ponderous—spanning 10–100 light-years and containing up to 100,000 times the mass of the

TABLE 3.1. Abridged Listing of Interstellar Molecules as Observed at Centimeter, Millimeter, and Sub-millimeter Wavelengths

2 atoms

Hydrogen (H_2)	Deuterized Hydrogen (HD)	Hydroxyl Radical (OH)	Silicon Monoxide (SiO)	Sulfur Monoxide (SO)
Nitric Oxide (NO)	Nitrogen Sulfide (NS)	Silicon Sulfide (SiS)	Methylidyne Radical (CH)	Cyanogen Radical (CN)
Carbon Monoxide (CO)	Carbon (C_2)	Carbon Monosulfide (CS)		

3 atoms

Water (H_2O)	Heavy Water (HDO)	Hydrogen Sulfide (H_2S)	Sulfur Dioxide (SO_2)	Nitroxyl (HNO)
Ethynyl Radical (C_2H)	Hydrogen Cyanide (HCN)	Deuterium Cyanide (DCN)	Formyl Radical (HCO)	Dicarbon Monoxide (C_2O)
Dicarbon Sulfide (C_2S)				

4 atoms

Ammonia (NH_3)	Formaldehyde (H_2CO)	Isocyanic Acid (HNCO)	Thioformaldehyde (H_2CS)	Acetylene (HC_2H)
Isothiocyanic Acid (HNCS)	Propynylidyne (C_3H)	Cyanoethynyl (C_3N)	Tricarbon Monoxide (C_3O)	Tricarbon Sulfide (C_3S)

5 atoms

Methane (CH_4)	Cyanamide (H_2HCN)	Formic Acid (HCOOH)	Cyanoacetylene (HC_3N)	Cyanomethyl (CH_2CN)

TABLE 3.1. (*Continued*)

5 atoms (*continued*)

Ketene (CH_2CO)	Methylenimine (CH_2NH)	Cyclopropenylidene (C_3H_2)

6 atoms

Butadinyl (C_4H)	Methyl Cyanide (CH_3CN)	Pentynylidyne (C_5H)
Methyl Alcohol (CH_3OH)	Methyl Mercaptan (CH_3SH)	
Formamide ($HCONH_2$)		

7 atoms

Methylamine (NH_2CH_3)	Methylacetylene (CH_3C_2H)	Hexatrinyl (C_6H)
Acetaldehyde (CH_3CHO)		Cyanodiacetylene (HC_5N)
Vinyl Cyanide (CH_2CHCN)		

8 atoms

Methyl Formate ($HCOOCH_3$)	Acetic Acid (CH_3COOH)	Glycoaldehyde ($H_2COHCHO$)

9 atoms

Methyl Cyanoacetylene (CH_3C_3N)	Dimethyl Ether (CH_3OCH_3)	Cyanohexatriyne (HC_7N)
Ethyl Alcohol (CH_3CH_2OH)		

10 atoms

Ethyl Cyanide (CH_3CH_2CN)		

11 atoms

Cyanooctatetrayne (HC_9N)		

12 atoms

Glycine (unconfirmed) (NH_2CH_2COOH)	Benzene (C_6H_6)	

13 atoms

Cyanopentaacetylene ($HC_{11}N$)	Methyl Diacetylene (CH_3C_4H)	

24 atoms

		Anthracene (PAH) ($C_{14}H_{10}$)

For more complete (and informative) listings, see http://www.astro.uni-koeln.de/cdms/molecules, and http://en.wikipedia.org/wiki/List_of _molecules_in_interstellar_space.

Sun. They are incredibly cold, with characteristic temperatures of only 5–20 degrees above absolute zero (where zero degrees Kelvin corresponds to -273 degrees Celsius). They are also redolent with organic molecules and microscopic grains of dust—the latter attribute making them dark in visible-light mappings. Indeed, a close match can be made between the overall distribution of CO emission and the sinuous rifts and blobs of darkness that E. E. Barnard photographed in the early twentieth century.

Because the CO molecule emits strongly at a discrete wavelength, the received wavelengths can be measured and interpreted according to the Doppler effect. By adopting a rotation curve for the Galaxy, as shown in figure 3.3, astronomers can then translate the observed galactic longitudes and radial velocities of the CO emission into distances. The resulting radial distributions of molecular clouds and associated star-forming activity have provided us with our first notion of the Galaxy's nebular architecture (see figure 3.5). We will revisit in chapter 9 the many issues that arise when using these spectral-line data to develop face-on maps, where the bugaboo of estimating distances once again rears its vexing head.

The Sub-millimeter Sky

The last frontier in mapping the Milky Way is at sub-millimeter wavelengths. This region of the electromagnetic spectrum spans wavelengths of several hundred micrometers (microns) to 1 millimeter. The Earth's atmosphere is especially adept at blocking radiation in this regime, as water and other abundant atmospheric molecules have many absorptive transitions that block the light from beyond. Some progress has been made, however, at observatories atop the European Alps (near the Matterhorn) and on the 13,700-foot Mauna Kea volcano in Hawaii. At these rarified altitudes, 90 percent of the atmosphere's water vapor lies below the telescope, and the sky opens up to sub-millimeter observing in particular spectral "windows." Oxygen levels are also reduced by 40 percent, leading many mountain-top visitors (including the author!) to experience the weirdness of intoxication without having ingested any

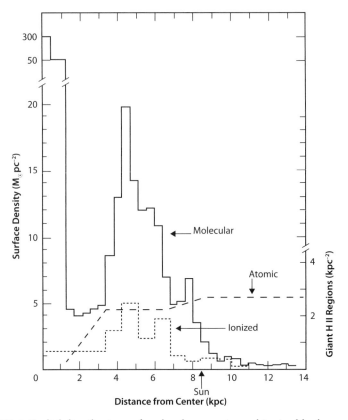

FIGURE 3.5. Radial distributions of molecular, atomic, and ionized hydrogen in the Milky Way. The molecular hydrogen (H_2) gas is traced by emission from the much brighter carbon monoxide (CO) molecule. The CO-emitting molecular gas is concentrated toward the galactic center and in a ring—or tightly wound spiral-arm pattern—located halfway out to the Sun's orbit. Radio observations of the ionized gas indicate that the so-called molecular ring is ablaze with rampant star formation. By contrast, the atomic hydrogen features a central "hole" and a flat distribution that extends well beyond the Sun's orbit. [Adapted from *The Fullness of Space* by Gareth Wynn-Williams, Cambridge, UK: Cambridge University Press (1992)]

alcohol. Those intrepid astronomers who have acclimatized to the severe oxygen deprivation at high altitude have gathered vital data on the coldest, most pregnant knots of star-forming gas. The dust grains in these gaseous blobs betray the presence of incipient protostars within, as their temperatures (of about 20 K) coincide with peak emissivities at sub-millimeter wavelengths (see figure 3.6).

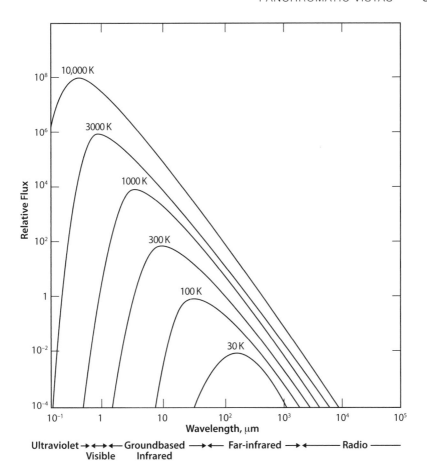

FIGURE 3.6. Spectral energy distributions of idealized thermal sources as a function of wavelength. Like the walls of pottery kilns, these so-called black bodies are in perfect equilibrium with the radiation that they are absorbing and emitting. The consequence is a spectral energy distribution (SED) that relies solely on the source's temperature. As the temperature increases, the wavelength of peak emission shortens, while the luminosity per unit area (surface flux) greatly increases. Dense opaque objects, such as the metallic filament of a light bulb or the heating element of a stove at temperatures of about 2,000K, provide close approximations to the black-body ideal. Perhaps surprisingly, the gaseous "surfaces" of stars with temperatures ranging from 2,000K to 50,000K, the human body at 310K, and the dusty clumps of star-forming gas at 10K–100K all radiate with SEDs similar to those plotted here. [Adapted from G. G. Fazio, in *Frontiers of Astrophysics*, ed. E. H. Avrett, Cambridge, MA: Harvard University Press (1976), p. 206. Based on B. Neugebauer and E. E. Becklin, "The Brightest Infrared Sources," in *Scientific American* 228 (1973): 28–40]

In May 2009, the Herschel Space Observatory was launched from French Guiana and into interplanetary space. This European-sponsored sub-millimeter observatory houses the largest telescope of any kind to be launched into space. It promises to open up the sub-millimeter spectral window as never before, reaping data on nearby motes of emergent starbirth along with the youthful outbursts of galaxies billions of light-years away. To follow the latest developments of this pioneering space observatory, see http://www .herschel.caltech.edu.

The Infrared Sky

Of all the spectral regions, the infrared regime is the richest and most complex. At *far-infrared* wavelengths (50–200 microns) the sky is awash in luminescence (see plate 6). Most of this emission arises from microscopic grains of dust that have been irradiated and warmed to several tens of degrees Kelvin by nearby stars. Some of this thermal emission is from discrete clouds, indicating the presence of newborn stars that are enshrouded with dust. Not surprisingly, these stellar hotbeds are located along the galactic midplane, where most of the star-forming molecular gas has settled. However, much of the far-infrared sky is gauzy and diffuse—as if an artist had applied a thin wash of paint across the firmament. The possible relationship between this lower-density medium and the denser material along the galactic midplane continues to puzzle astronomers.

By considering the *relative* enhancements and depletions of far-infrared emission, one can make maps that reveal exquisite structuring at every galactic latitude (see figure 3.7). This form of image processing through judicious *spatial filtering* is a bit of an artifice. However, it brings out details that otherwise would be lost in the increasing glare toward the galactic midplane. It is now possible to perceive shell fragments, sheets, loops, and filaments of dusty nebulosity on all size scales. Indeed, the richness of structure is far too much to readily comprehend. What are these emitting apparitions really tracing? And what agents are responsible for having sculpted them?

We are fairly confident that at wavelengths of 50 to 200 microns, the infrared emission is a reasonable proxy for most of the dust in the Milky Way. The dust, in turn, is thought to be well-mixed with the gas—whether it be in molecular, atomic, or ionized form. That means these complex scenes can be regarded as fair representations of the ornately structured interstellar medium in our Galaxy. We are left agog at the *fractal* character of the nebulosity, where the view on small scales is comparable to that on much larger scales. These sorts of considerations can help us to constrain what might be the major movers and shapers of the interstellar medium. Our best guess is a combination of large- and small-scale energizers.

Besides irradiating and warming the nebular dust, nearby stars can also shape their surrounding nebulae. Massive hot stars can photo-ionize and photo-evaporate their natal nebulae. They can also blow strong winds that can move things around. Finally, they can die in spectacular supernova explosions that inflate giant bubbles of plowed-up gas. Lower-mass stars like the Sun are at their windiest in their infancy and bloated old age. Though less powerful than their massive cousins, they are more numerous and so can collectively pack a punch. Stellar powering of the interstellar medium is certainly important on scales of 0.1 to 10 light-years, and may be influential on scales of 100 light-years or more. On even larger scales, the dynamics of the swirling galaxy and of its magnetic fields may play some role. And on all scales, turbulent motions may help to structure the interstellar medium into the froth that is evident in far-infrared mappings. To learn more, we can once again resort to insights at other wavelengths.

At *mid-infrared* wavelengths (5–50 microns), much of the emission is thought to arise from organic molecules. These so-called polycyclic aromatic hydrocarbons (PAHs) are aggregates of many benzene ring molecules (C_6H_6) in the form of flexible sheets. The bending and stretching of these molecular sheets are the likely mechanisms for producing the spectral emission features that have

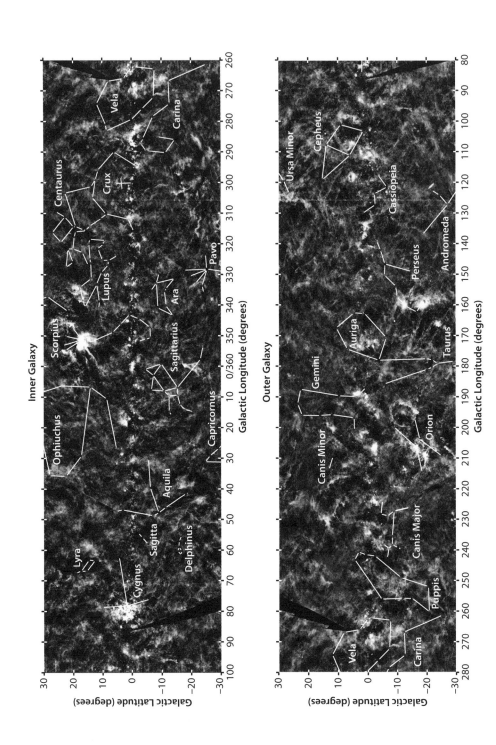

Inner Galaxy

Galactic Longitude (degrees)

Galactic Latitude (degrees)

Lyra
Cygnus
Sagitta
Delphinus
Aquila
Ophiuchus
Scorpius
Lupus
Centaurus
Crux
Vela
Carina
Pavo
Ara
Sagittarius
Capricornus

Outer Galaxy

Galactic Longitude (degrees)

Galactic Latitude (degrees)

Ursa Minor
Cepheus
Cassiopeia
Andromeda
Perseus
Auriga
Taurus
Gemini
Orion
Canis Minor
Canis Major
Puppis
Vela
Carina

been observed at mid-infrared wavelengths. The Spitzer Space Telescope, launched in 2003, has revealed that the Milky Way and other star-forming galaxies are redolent with these emitting PAHs. Again, the emission is arranged into fabulous forms—much of which is associated with newly forged stellar populations. We will examine these regions more closely in chapter 6, when we focus on the nebular birthplaces of stars. On Earth, PAHs are as familiar to us as the mouth-watering aromas of a barbecued hamburger, the sweetly acrid odors of burning tobacco, and the choking fumes behind a diesel bus. If we had big enough nostrils, what would our home galaxy smell like?

At *near-infrared* wavelengths (1–5 microns), the nebular Milky Way dims to insignificance, as the stellar component of our Galaxy brightens to greatest prominence (see plate 6). Compared to visible light, the near-infrared emission is 2–10 times less affected by intervening dust. That means it provides a better representation of the overall stellar distribution in the Milky Way. Using all-sky maps of the near-infrared emission, astronomers have discovered that the stellar disk of the Milky Way is warped like a stylish fedora and that the central bulge is in the form of an ob-

---

FIGURE 3.7. Maps of far-infrared emission in the Milky Way as recorded by the Infrared Astronomy Satellite (IRAS) in the 1980s and subsequently processed to highlight the multiscale structuring. The nature and origin of these baroque scenes remain poorly understood. A cursory comparison with figure 1.2 and plates 2–5 (which show the same regions at visible wavelengths) reveals dramatic differences in the emitting structures. (*Top*) Structuring of far-infrared emitting dust in the inner galaxy. Galactic longitude increases from left to right—from 260° through 360° (0°, the galactic center) to 100°. Galactic latitude ranges from bottom to top, from -30° through 0° (the galactic midplane) to +30°. (*Bottom*) Structuring of far-emitting dust in the outer galaxy. Galactic longitude increases from left to right—from 80° through 180° (the galactic anticenter) to 280°. [Composed by W. H. Waller and D. Lampert from data obtained by the Infrared Astronomy Satellite (IRAS) and processed by the Infrared Analysis and Processing Center (IPAC), Jet Propulsion Laboratory, Caltech, NASA. These data were accessed using the *SkyView* interactive archive at NASA's Goddard Space Flight Center (see http://skyview.gsfc.nasa.gov). IRAS data reference: Wheelock, et al. 1991, IRAS Sky Survey Atlas Explanatory Supplement. Data from the Infrared Astronomy Satellite (IRAS)]

long bar. Recent estimates of these structural elements are shown in figure 3.8.

The Visible Sky

As featured in plates 1–5, long-exposure imaging of the visible sky reveals a hodge-podge of starlight, silhouetting dark clouds, and roseate nebulae. Our visible perspective of the galactic disk is biased to the nearest 5,000 light-years or so, as the dust clouds obscure most of what lies beyond. Nevertheless, within this purview there are plenty of wondrous celestial objects to keep astronomers up at night. These include star-forming nebulae, stars of all kinds, planetary systems around these stars, planetary nebulae (an arcane and incorrect name for the fluorescent nebulae ejected by aged out-gassing stars), supernova remnants, compact stellar remnants (white dwarfs, neutron stars, and black holes), interacting binary star systems, rich star clusters, nova outbursts, and other surprises. Indeed, most of what we know about stars and stellar remnants in the Milky Way has been obtained through observations at "optical" wavelengths (0.3–1.0 microns, or equivalently 3,000 to 10,000 Angstroms—where an Angstrom is 10^{-10} meters).[2]

Ionized nebulae in the Milky Way are especially prominent at optical wavelengths. As excited electrons cascade down to lower energy levels within their hosting ions (atoms stripped of one or more electrons), they radiate spectral-line emission at wavelengths corresponding to their ordained quantum jumps. Ions of oxygen, nitrogen, and sulfur are commonly observed in nebulae that have been excited to fluoresce by the nearby presence of hot ultraviolet-bright stars or by the action of shock waves from some nearby explosion or outburst. For example, we often see emission from singly ionized oxygen ([OII]), sulfur ([SII]), and nitrogen ([NII]), the brackets indicating that the electronic jumps are of low probability (i.e., "forbidden") yet nonetheless occur in the relative vacuum of space—where de-exciting collisions are rare. In hotter and

FIGURE 3.8. The overall distribution of long-lived stars in the disk and central bulge of the Milky Way, as surmised from the near-infrared mapping survey by the Cosmic Background Explorer (COBE). (*Top*) Face-on perspective with the Sun denoted by a white dot. The bulge is configured as a bar pointing 15 degrees from the Sun. (*Bottom*) Edge-on perspective showing a warp in the outer stellar disk. [Courtesy of H. Freudenreich, Diffuse Infrared Background Explorer (DIRBE), COBE, NASA, with reference to H. Freudenreich, "Deconstructing the Milky Way Galaxy," *American Scientist* 87 (1999): 418]

more energized nebulae, emission from twice- and thrice-ionized versions of these species are common. The degree of ionization—as perceived by the characteristic suite of spectral emission lines—provides astronomers with vital clues to the energetic processes that are affecting these nebulae.

The brightest optical tracer of ionized gas in the Milky Way is the hydrogen "alpha" emission that occurs when a hydrogen ion (a bare proton) recaptures a free electron. As the bound electron cascades to lower energy levels, it emits the complete spectrum of hydrogen emission lines. The jump from the third to second energy level produces the familiar ruby-red H-alpha emission at 6,563 Angstroms. This can be seen throughout the Milky Way in the form of brilliant star-forming knots and vast expanding shells (see figure 3.9). We will revisit these so-called *HII regions* elsewhere in this book, where stellar-nebular feedback processes are explored in depth.

The Ultraviolet Sky

Our atmosphere's ozone layer effectively blocks most ultraviolet light from penetrating to the Earth's surface. That is good news for those who do not wish to experience terminal sunburns. However, it makes ultraviolet astronomy the exclusive domain of balloon-, rocket-, and space-borne telescopes. This spectral domain has been

FIGURE 3.9. Ionized gas pervades the galactic disk and extends to great distances above and below the galactic midplane. (*Top*) All-sky rendering of diffuse H-alpha emission from ionized hydrogen in the Milky Way. The excited hydrogen traces the environmental effects of intense star-forming activity. (*Middle*) Mosaic of diffuse far-ultraviolet continuum emission from 80% of the sky, as recorded by the Korean-American SPEAR/FIMS spacecraft. The continuum mosaic traces interstellar dust that has scattered UV light from hot stars. (*Bottom*) The Carbon IV mosaic by Spear/FIMS traces UV spectral-line emission from three-times ionized carbon at very high temperatures indicative of shock heating from supernova explosions. The most prominent CIV-emitting source is the Vela supernova remnant to the right of center. [(*Top*) Made from the VTSS/SHASSA/WHAM H-alpha surveys of the sky, with reference to D. P. Finkbeiner, "A Full-Sky H-alpha Template for Microwave Foreground Prediction," *Astrophysical*

Ionized Hydrogen

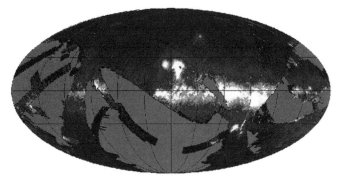

Ultraviolet Continuum

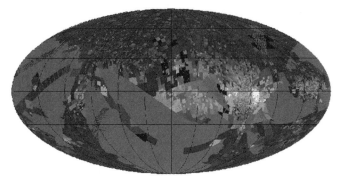

Carbon IV

*Journal Supplement*146 (2003): 407. Data accessed by W. H. Waller from the SkyView interactive archive at NASA's Goddard Space Flight Center (see http://skyview.gsfc.nasa.gov). (*Middle and Bottom*) SPEAR/FIMS mission, NASA/MOST, with reference to J. Edelstein, K. W. Min, W. Han et al., "The Spear / FIMS Mission," *Astrophysical Journal* 644 (2006): L153. Images and data are archived at http://spear.ssl.berkeley.edu/archive.html]

hard-won, with significant results arising in just the last twenty-five years.

Having wavelengths of 100–3,000 Angstroms, ultraviolet radiation is also readily absorbed by the many—even smaller—grains of dust that permeate the Milky Way. For UV radiation with wavelengths less than 912 Angstroms, the hydrogen gas itself becomes an efficient absorber, the H atom's sole electron grabbing any extreme-UV photon in order to make its escape. It is therefore not surprising that all-sky surveys of ultraviolet continuum emission are underwhelming in their content. The resulting maps show a few hundred hot stars distributed across the sky. Many of them turn out to be white dwarf stellar remnants located within only 100 light-years of the Sun. Away from the dusty disk of the Milky Way, the obscuration is much less, and more distant objects can be observed at ultraviolet wavelengths. For example, UV imaging by a telescope that operated aboard the Space Shuttle *Columbia* recorded collectives of hot stars that reside in globular clusters of the galactic halo—some 30,000 light-years away. These ancient stars are breathing their last gasps. Having recently shed their outer layers, they lay mostly spent, their hot inner cores fully exposed and glowing brightly in the UV (see figure 3.10).

Despite the limiting effects of dust, the UV sky is rich with spectral-line emission from atoms that have lost three or more of their electrons. The energetics required to strip these atoms of their outer electrons are impressive, indicating temperatures of up to several hundred thousand degrees. These sorts of temperatures can be found in the shocked shells of supernova remnants, where the gas is cooling from millions of degrees to several thousand degrees. A recent effort to scan UV line emission across the sky has resulted in partial maps of the diffuse emission from three times ionized carbon ([CIV]), five times ionized oxygen ([OVI]), and other highly ionized species (see figure 3.9). These pioneering maps portend more revealing UV panoramas of the Galaxy in the not-too-distant future.

Meanwhile, UV spectroscopy of the many brilliant and remote quasars that pepper the sky have yielded spectral plots that often

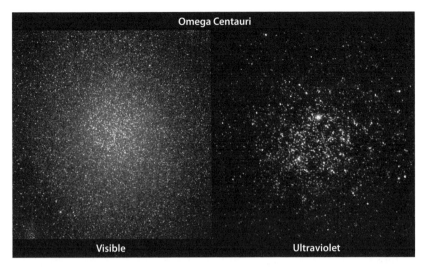

FIGURE 3.10. Visible and ultraviolet views of the massive globular cluster Omega Centauri. This cluster contains more than a million stars, all of which were born about 12 billion years ago. The visible image (*left*) shows the many stars that have bloated into their red-giant phase. The ultraviolet image (*right*), recorded by the Space Shuttle-borne Ultraviolet Imaging Telescope, shows those stars that have evolved further, having shed their outer layers and exposed their hotter interiors. [From *Beyond the Blue—Greatest Hits of the Ultraviolet Imaging Telescope*, Slide set and booklet by W. H. Waller and J. Offenberg, San Francisco: Astronomical Society of the Pacific (1996). Excerpts archived by the Space Telescope Science Institute, NASA at http://archive.stsci.edu/uit/project/Astro1/Astro1_pictures.html]

include *absorption* features produced by extremely hot clouds of gas in the "foreground" halo of our Galaxy. These absorption-line features enable highly accurate measures of the clouds' temperatures, densities, and kinematics. We now recognize that the galactic halo is run through with tenuous clouds of torrid gas. The origin of these clouds is uncertain, but many astronomers look toward recent supernovae in the Galaxy's disk having exhausted their gaseous outbursts into the halo. Astronomers further imagine this hot gas slowly cooling to lower temperatures and eventually raining back down onto the disk in what has been described as a *galactic fountain*.

The X-ray Sky

X-radiation vibrates at fantastically small oscillatory wavelengths of 0.1–100 Angstroms—the equivalent size of individual atoms and molecules. In the dentist's office, it is produced by accelerating electrons across a strong voltage difference and then smashing these electrons onto a metallic "target." The sudden deceleration of the electrons releases energy in the form of X-ray photons that can pierce through our soft tissues and reveal regions of dental decay. In space, X-radiation is produced by similarly energetic processes—either through acceleration of charges by explosive and magnetically intense events or by heating gas to millions of degrees Kelvin. None of this cosmic X-radiation makes it through the Earth's atmosphere to the surface, and so X-ray astronomy has been the exclusive province of space-borne instruments.

X-ray astronomers rarely speak in terms of oscillatory wavelengths or frequencies. Instead, they consider the energies of the individual X-ray photons. Their favored unit of energy is the electron-Volt (eV), where one eV amounts to the energy given to one electron as it accelerates across a voltage difference of one Volt. The photons that our eyes detect and our brains interpret according to color have energies of a few eV. The ionizing UV photons that produce HII regions have energies exceeding 13.6 eV. X-ray astronomers deal with photons having energies of hundreds to millions of eV.

Recent mappings of the X-ray sky are often color-coded according to the detected photon energies—with blue denoting the highest energies and red the lowest energies (see plate 6). These X-ray vistas reveal discrete regions where stars have exploded, along with more diffuse emission of uncertain origin. Two bright regions especially stand out—the Cygnus Loop supernova remnant to the left, and the Puppis/Vela supernova remnant complex to the right. The preponderance of diffuse emission from the general direction of the galactic center continues to elicit debate. Is it tracing some huge outburst from thousands of recent supernova explosions near the galactic center, explosions from hundreds of supernovae

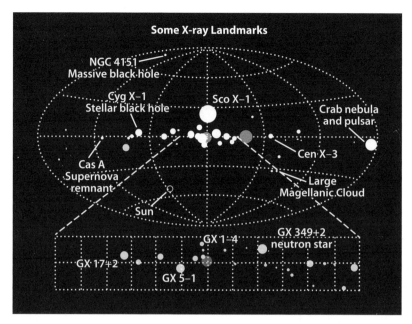

Some X-ray Landmarks

NGC 4151
Massive black hole

Cyg X–1
Stellar black hole

Sco X–1

Crab nebula
and pulsar

Cen X–3

Cas A
Supernova
remnant

Large
Magellanic Cloud

Sun

GX 349+2
neutron star

GX 1–4

GX 17+2

GX 5–1

FIGURE 3.11. Still from a movie of varying X-ray sources in the sky with closeup of the inner 60 degrees below. The animation can be accessed from the Rossi X-ray Timing Explorer website at http://xte.mit.edu. [From D. A. Smith, M. Muno, A. M. Levine, R. Remillard, H. Bradt, and the All-Sky Monitor (ASM)—Rossi X-ray Transient Explorer (XTE) Science Team, MIT and NASA's Goddard Space Flight Center. Movie archived at http://xte.mit.edu]

in the molecular ring interior to the Sun's orbit, or something of more local and less powerful origin? Recent analyses place the emitting gas fairly near the galactic center, thus indicating a humongous bipolar outflow that involved some 10,000 supernovae having detonated about 15 million years ago.

Besides the diffuse emission, the X-ray sky is peppered with discrete sources that vary dramatically over time (see figure 3.11). The Rossi X-ray Transient Explorer (RXTE) was designed to record these histrionics over days, weeks, months, and years. Between February 1996 and December 1998, the RXTE monitored the entire sky. The resulting animation of emitting sources features outbursts from pulsars, neutron stars, and likely black holes. These so-called compact stellar objects are thought to be in close binary systems, where they are feeding off their companion stars. The gases from the victimized stars are accelerated to extreme speeds by the phenomenal gravity of the compact companions. As the gases crash onto the accretion disks that surround these bizarre worlds, they betray their violent ends in the language of X-rays.

The Gamma-ray Sky

Like X-radiation, gamma radiation is studied in terms of the photon energies that are involved. Here the favored units are in millions to billions of electron Volts (MeV to GeV). Indeed, gamma-ray photons are so energetic (and rare) that it is tempting to give each and every one of them a name. Detecting them involves tracking their interactions with matter in some form or other. Some detectors record the optical light that is released when the gamma rays interact with crystals in the instrument. Other detectors use the Earth's atmosphere itself as the luminescent medium. Still others capture the high-energy photons in "spark chambers" full of gas which then produce sprays of electrons and other particles. The direction and energy of the streaming electrons can then be recorded by their interactions with wire grids inside the chamber.

The first gamma-ray observatory to map the entire Milky Way had four of these elaborate—and heavy—detecting devices. The 17,000-kilogram *Compton Gamma-Ray Observatory* was lugged up to orbit by the Space Shuttle *Atlantis* in April 1991. The first of the "Great Observatories" (including the Hubble, Chandra, and Spitzer observatories), the CGRO operated for nine years. When it lost one of its stabilizing gyroscopes, it was purposefully de-orbited and allowed to crash into the Pacific Ocean in June 2000.

Of the four gamma-ray instruments onboard the CGRO, the Energetic Gamma Ray Observatory (EGRET) mapped the gamma-ray sky at the highest energies (20–30 GeV). The resulting all-sky panorama reveals some of the lowest-energy matter in the Galaxy (see plate 6). This may seem contradictory, but it turns out that gamma rays in this energy range are created in abundance, when cosmic rays (relativistic particles) interact with the gaseous matter that pervades the Milky Way. Because the cosmic rays are not choosy in their interactions with the various gaseous phases, they end up "illuminating" the denser concentrations of cold molecular and cool atomic gas most effectively. Besides tracing the overall interstellar medium, the CGRO has also revealed energetic plumes of diffuse gamma-ray emission emanating from the Galaxy's nucleus—thus confirming similar X-ray detections and calling into question the origins of all these nuclear theatrics.

Many marvels of the gamma-ray sky remain to be discovered. Fortunately, new instruments are being developed that will greatly improve upon the angular resolutions and time sensitivities that can be achieved. With these improvements, entirely new categories of cosmic high-energy phenomena may be revealed. Many astronomers have their hopes pinned on the *Fermi Gamma-ray Space Telescope* that was launched in June 2008. The latest news on this major new observatory can be found at http://fermi.gsfc.nasa.gov/.

CHAPTER 4

· · ● ● · ·

NEIGHBORS OF THE SUN

There's no place like home!
There's no place like home!
There's no place like home!

—Dorothy in *The Wonderful Wizard of Oz*,
by L. Frank Baum (1900)

THE GREAT SPIRIT must have been especially magnanimous when creating the Milky Way Galaxy.[1] Spanning more than 100,000 light-years and containing more than 100 billion stars, the Milky Way rules over the Local Group of galaxies in a shared arrangement with Messier 31—the great Nebula in Andromeda. Together, these two giant spiral galaxies account for more than 85 percent of the luminous matter in the Local Group—with the other forty-odd galaxies swarming about their two massive hosts like bees around twin hives.

We have learned of these amazing proportions through the classic astronomical task of determining distances to sundry luminous sources in the night sky. This same gauging activity has also helped us to determine which objects are closest to us—within what is commonly called the solar neighborhood. Indeed, our fathoming of distances to stars in the solar neighborhood comprises the key stepping-stone for determining distances to everything else in the universe. From these first steps into the cosmos, we have discovered a strange assortment of stars and planetary systems which, for the most part, bear little resemblance to our Sun and Solar System.

A Stellar Census

Although astronomers differ on its precise dimensions, the solar neighborhood measures approximately 100 light-years in radius. To put things in perspective, this patch of galactic real estate occupies less than 1/250,000 of the area spanned by the Milky Way's disk. The period at the end of this sentence marks a similar fraction of the page that you are reading.

Within the solar neighborhood, distances to stars can be reckoned through the direct technique of heliocentric parallax. Stellar positions are recorded with respect to more distant "background" stars. As the Earth orbits around the Sun, these positions are seen to shift ever so slightly (see figure 4.1). Over a half-year, our observing platform (the Earth) travels from one side of the Sun to the opposite side, and the parallactic shift reaches its maximum. In the next half-year, the shift declines back to zero. This parallactic cycle is repeated every Earth year.

The maximum angular displacement of the foreground star with respect to the background stars is handily related to the distance of the star by the formula

distance (parsecs) = 1 / parallax (arcseconds),

where the parallax angle is that measured over a quarter-year, 1 arcsecond is 1/3,600 of a degree, and 1 parsec is 3.26 light-years (or 30.8×10^{12} km). Here, the parsec unit of distance is seen to be operationally defined as a "parallax arcsecond," whose contraction is "parsec." Because of this functional basis, astronomers tend to favor "parsec" over "light-year" when quantifying distances to celestial objects. In this book, however, we will often stick with "light-year" as being the more intuitive unit of expressing cosmic distances.[2]

Current space-borne instruments can discern parallactic shifts as small as 1 milli-arcsecond (0.001 arcsec). According to the previous formula, the corresponding limiting distance for this reckon-

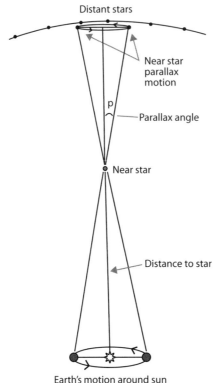

Distant stars

Near star
parallax
motion

p

Parallax angle

Near star

Distance to star

Earth's motion around sun

FIGURE 4.1. Schematic depiction of the parallactic shifting in angular position that a foreground star undergoes with respect to more distant "background" stars. The observed shifting is due to the changing location of the Earth in its orbit around the Sun. This rendering of parallactic shifting has been greatly exaggerated to show the pertinent angles and distances. The measured parallaxes of even the nearest stars are significantly less than 1 arcsecond. [Illustration by L. Slingluff from multiple sources]

ing technique is 1,000 parsecs or 3,260 light-years. Residents of the solar neighborhood all lie well within this range. Moreover, each and every star within the solar neighborhood is near enough to be identified and characterized—no matter how faint it is. Most of the stars turn out to be incredibly faint, requiring large telescopes to study them in any detail.

Let us begin with our closest stellar neighbor—the Alpha Centauri triple-star system. Observers in the southern hemisphere are blessed with brilliant views of Alpha Centauri during the months of August through December. Blazing forth at -0.27 magnitude,[3] the yellowish Alpha Centauri system gilds the night sky along with bluish Beta Centauri and the exquisite Southern Cross. Telescopic observations reveal a triple-star arrangement, with Alpha Cen A shining brightest at 0.01 mag, orange Alpha Cen B about 3 times

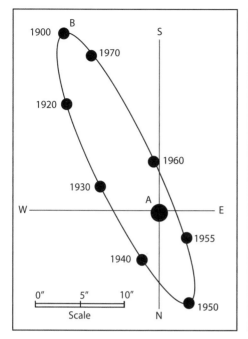
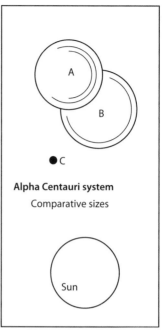

FIGURE 4.2. The Alpha Centauri system consists of Sun-like Alpha Centauri A along with cooler and dimmer Alpha Centauri B. These two stars delineate 80-year orbits about their mutual center of mass. We observe this stellar dance from the reference frame of Alpha Cen A, with Alpha Cen B appearing to swing around its more ponderous mate in an orbit that is of similar size as Neptune's orbit around the Sun but is far more elongated. Proxima Centauri (Alpha Centauri C) can be found about 2 degrees from these two entwined stars, where it is thought to be circumnavigating them in a million-year orbit—or perhaps a hyperbolic flyby. It is currently 0.24 light-years closer to us than the rest of the system, thus qualifying it as the most "proximate" of stars. [Adapted from *Burnham's Celestial Handbook*, by R. Burnham, Jr., New York: Dover Publications (1978), p. 550]

fainter at 1.34 mag, and remotely orbiting Proxima Centauri with a ruddy glow of 11.09 mag that translates to being 17,000 times dimmer than Alpha Cen A (see figure 4.2).

The measured parallax of this celestial beacon is 0.747 arcsecond, and so the corresponding distance is 1.338 parsecs or 4.36 light-years. At this distance, the perceived brightness of Alpha Cen A translates to an absolute luminosity that is remarkably close to that of the Sun. Its yellow color and detailed spectrum are also very Sun-like, yielding a surface temperature of 5,800 Kelvins. Given the plethora of dim bulbs throughout the rest of the solar neighborhood, one cannot help but be amazed that the closest stellar system to the Sun contains a much brighter star that could masquerade as the Sun's sister. Does this happenstance indicate a common origin? Probably not, alas. One need only consider the incredible distances that are involved.

Were the Sun reduced to the size of a grapefruit (about 10 cm) and placed on the Mall in Washington, DC, the Earth would be but a mustard seed about 11 meters away. Jupiter would be a more substantial blueberry 55 meters from the solar grapefruit. Pluto and the other "dwarf" planets that occupy the icy belt of debris beyond the orbit of Neptune would be the equivalent of salt grains a full 420 meters from their source of citric power. All of this scaled-down planetary activity could easily fit within the National Mall's grassy expanse. By contrast, Alpha Centauri would be another grapefruit a whopping 2,800 km away—sitting by some lonely roadside near Albuquerque, New Mexico. These sorts of uncanny separations make any past and future associations between Alpha Centauri and the Sun highly unlikely.

In truth, we have no idea which stars in the Galaxy are related to us by birth. Far too much time has elapsed since our stellar family emerged from the same placental cloud of molecular gas some 4.6 billion years ago. We have long since dispersed and gone our own ways—destined to be strays among strays. Perhaps someday, astronomers will have a sufficiently detailed understanding of stellar ages, chemical compositions, and motions throughout the Milky Way. Only then will we be able to identify our stellar kinfolk and their planetary progeny. For now, we had best content ourselves with knowing that most of the stars in the solar neighborhood were probably born at similar distances from the galactic center and are endowed with similar chemical compositions. Therefore, the solar neighborhood provides a representative sampling of the stars that make up the disk some 20,000–30,000 light-years from the galactic center.

A cursory examination of figure 4.3 shows that most of our closest stellar neighbors are puny red dwarfs like Proxima Cen-

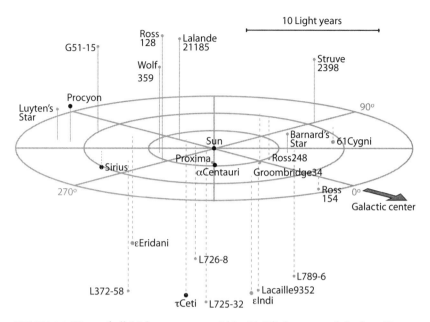

FIGURE 4.3. View of all 33 known stars within 12.5 light-years of the Sun. Despite their proximity, most of these stars are very dim, requiring powerful telescopes to be observed. [Adapted from http://www.atlasoftheuniverse.com/, by R. Powell, licensed under a Creative Commons Attribution-ShareAlike 2.5 License]

tauri. The Sun and Alpha Centauri A are the exceptions rather than the rule. The other notable outliers include the brilliant A-type star Sirius and the F-type star Procyon—my all-time favorite stellar name. The remaining stars within 12.5 light-years of the Sun have luminosities ranging from 1/10,000 to 4/10 that of the Sun with estimated masses of 6/100 to 7/10 solar masses.

Expanding our view to a radius of 50 light-years increases our sample to about 2,000 stars. This is enough to derive some robust statistics on the types of stars that roam our part of the Galaxy. Figure 4.4 shows the breakdown in terms of luminosities and temperatures. Clearly, the dim and cool red M-type stars prevail. As in so many other populations (mice vs. elephants, herring vs. swordfish, ferns vs. redwoods, etc.), the small and meek are the most ubiquitous.

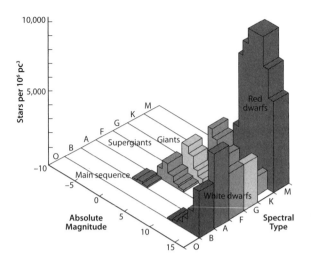

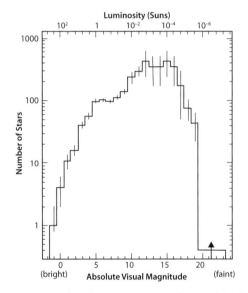

FIGURE 4.4. (*Top*) Proportions of stellar types in the solar neighborhood as plotted on a Hertzsprung-Russell diagram, where temperature increases from the cool M to hot O spectral types. The most numerous stars are the dim red M-type dwarfs along the Main Sequence, followed by the hotter but much smaller white dwarfs. (*Bottom*) The detailed frequency distribution of stellar luminosities also shows a broad peak at the low luminosities characteristic of red dwarfs. Moreover, it shows a bump at luminosities slightly dimmer than that of the Sun (M_V = 4.85 mags). This bump of orange main-sequence stars may indicate that a spate of star formation occurred in the Milky Way some 10-12 billion years ago. [(*Top*) From Seeds, *Horizons: Exploring the Universe (with The Sky CD- ROM and Info-Trac)*, 7E. © 2002 Wadsworth, a part of Cengage Learning, Inc. Reproduced by permission. www.cengage.com/permissions. (*Bottom*) Adapted from H. Jahreiss and R. Wielen, in *ESA SP-01: Proceedings of the Hipparcos-Venice '97 Symposium*, ed. M.A.C. Perryman and P. L. Bernacca, 402 (1997), 67]

A careful plotting of these stellar numbers indicates a bump at an absolute magnitude of 5.0 or so. This slight overabundance pertains to stars with luminosities half that of the Sun and total lifetimes of about 12 billion years. Could this be saying something? Perhaps a burst of star formation accompanied the early development of the galactic disk. We would then be seeing the stellar leftovers from that seminal event.

After the copious red dwarfs, the next most abundant species are the white dwarfs. Hot as Sirius but with diminutive sizes no greater than that of Earth, these cooling embers of once Sun-like stars trace the early history of the Milky Way. By counting the numbers of white dwarfs as a function of luminosity, astronomers have found a clear low-luminosity cutoff amounting to 1/100,000 that of the Sun. Because white dwarfs cool and dim at a well-known rate, the lowest-luminosity WDs should also be the oldest. The observed luminosity cutoff is thought to correspond to an age of about 8–10 billion years. This maximal white dwarf age provides a fairly secure lower limit to the age of the galactic disk itself. A more complete comparison of "chronometers" in the disk is presented in chapter 11, where various scenarios for the Milky Way's origin and evolution are explored.

The Nearest Exoplanetary Systems

The first extra-solar planet was found in 1992 by radio astronomers who observed the pulsar PSR 1252+12 to periodically vary in the timing of its millisecond pulses. This was interpreted as the consequence of the gravitational effect between the planet and the pulsar—causing the pulsar to orbit the mutual center of mass and thereby modulate its beeping signal as observed by us. The first planet to be found orbiting a "normal" main-sequence star was discovered in 1995 by Swiss astronomers who had made precise spectroscopic observations of 51 Virginis. As of this writing, more than 800 planets beyond the Solar System have been confirmed, several of which are in multiple exoplanetary systems. Figure 4.5 shows a selection of multiple exoplanetary systems with their respective arrangements around their host stars. The figure also shows the frequency distributions of orbital distances and planetary masses. Most planets are of Jupiter class and are found very close to their host stars.[4]

The observed preponderance of massive exoplanets inside the equivalent orbit of Mercury (0.4 AU) is a startling result that has many theoretical astrophysicists positing new theories for the birth and early inward migration of giant planets within circumstellar disks. However, such cozy relations between these planetary behemoths and their host stars may be the consequence of serious observational selection effects at work. It is well worth our while to understand these effects before jumping to any further conclusions.

Most (95 percent) of the confirmed exoplanet detections are indirect, involving spectroscopic observations of stars in the solar neighborhood over long periods of time. These observations are by necessity exquisitely precise, as they must detect the reflexive response of the star to the gravitational nudging of one or more planets. The wobbling motions of these stars amount to no more than a few tens of meters per second—typical highway speeds. By virtue of the Doppler effect, this minuscule oscillation translates to a tiny shifting of the star's spectroscopic absorption lines. The Doppler shifting in observed wavelength is of order 1/10,000 of an Angstrom, requiring extraordinarily high spectral resolutions and advanced analytic techniques.[5]

Despite these practical challenges, astronomers have devised powerful spectroscopes that can track the infinitesimal spectral-line variations as a star responds to its planetary perturber(s). Those stars with massive planets in tight orbits are gravitationally jostled the most, and so are most readily detected. Much of what we can see in figure 4.5 can be attributed to this strong observational selection effect.

A newer and very promising detection technique involves monitoring the central star's light output, as a planet passes in front of it. If the Sun was observed by some distant civilization while the Earth "transited" in front of it, the alien observers would detect a corresponding diminution of light amounting to the ratio of projected areas. The Earth has a diameter roughly 1/100 that of the

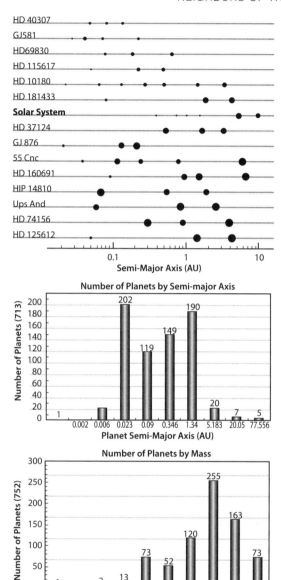

FIGURE 4.5. (*Top*) Multiple planetary systems that have been recently discovered show a wide range of sizes and distances from their host stars (where 1 AU is the Sun-Earth distance). (*Middle*) The frequency distribution of orbital distances (in AU) for planets in both single and multiple systems shows a bias toward relatively small orbits. (*Bottom*) The frequency distribution of planetary masses (in equivalent Jupiter masses) is dominated by giant planets. [(*Top*) Adapted from C. Lovis et al., "The HARPS Search for Southern Extra-solar Planets XXVIII . . . " *Astronomy and Astrophysics* 528 (2011): A112. (*Bottom*) Created by W. H. Waller from the interactive exoplanet catalog at http://exoplanet.eu/]

Sun, and so its projected area is $(1/100)^2$ or 1/10,000 that of the Sun. Detection of the Earth in transit would require monitoring the light output with accuracies significantly better than one part in 10,000 (or 0.01 percent).

Transiting Jupiters are much easier to detect, as they obscure a more substantial 1/100 of their host suns. Once again, we see an observational selection effect at work. Most of the transiting exoplanets that have been detected so far are of the giant Jupiter variety. The transiting exoplanets are also found to orbit very close to their central stars. This again makes sense, if you consider the likelihood of catching a star being eclipsed by one of its planets. Those planets closer in can have a larger range of orbital inclinations and still eclipse the star as seen by us. Some of these planetary transits have been successfully confirmed by amateur astronomers using commonly available instrumentation.

Spectroscopic observations of transiting planets have given us our closest views of these newfound worlds. In 2001, astronomers used the Hubble Space Telescope to obtain detailed spectra of the central star HD 209458, as its orbiting planet passed in and out of eclipse. They then subtracted the pure stellar spectrum from the eclipse spectrum, producing a residual spectrum of the planet's atmosphere. This amazing feat revealed the telltale signs of sodium in the giant planet's upper atmosphere. Subsequent ground-based observations of other transiting systems have revealed hot planetary atmospheres laden with molecular hydrogen, methane, and water vapor (steam), the high temperatures arising from the nearby star's scorching radiation.

In the coming decades, new space observatories will search for planetary transits around hundreds of thousands of nearby stars. Among the first to do so is the *Kepler* mission that was launched in March 2009. Over its 3.5-year lifetime, it has been monitoring 100,000 stars in a $10° \times 10°$ field of view that is situated between the constellations of Cygnus and Lyra. Being spaceborne, *Kepler* does not have to contend with the vagaries of observing through the Earth's atmosphere. It can therefore achieve photometric accuracies of 0.002 percent and so can begin to harvest a truer representation of large and small planets. Already, *Kepler* has identified more than 1,200 candidate exoplanets (it takes several transits to confirm planetary status), including planets with sizes similar to those of the ice giants Uranus and Neptune—and even smaller planets, what astronomers have dubbed "super-Earths." Several of these planets have been found at fairly comfortable distances from their host stars, where water could be present in liquid form.

Of course, the holy grail of planet finding is to directly image an Earth-like exoplanet. The practical requirements are incredibly daunting, equivalent to spotting a gnat buzzing around a spotlight on the Moon. NASA's *Terrestrial Planet Finder* mission has been charged with doing just that. This ambitious venture is in the initial stage of development with promising ideas for its design being prototyped and tested in selected laboratories. Meanwhile, the first direct images of giant exoplanets have been obtained recently with the Hubble Space Telescope and the ground-based Very Large Telescope. By blocking out the glare from the central stars and taking multiple images over time, astronomers have confirmed that the observed motes of light are in orbital motion around their host stars and hence are most likely bona-fide planets (see figure 4.6).

The Local Bubble

All of the stars and planetary systems in the solar neighborhood are drifting through an ornate filigree of tenuous gas and dust. This so-called Local Interstellar Medium (LISM) is what connects us with the rest of the Milky Way, making us part of a vast galactic ecosystem that continues to evolve and transmogrify. Our Sun and Solar System are not excluded, as we are currently passing through a small part of the filigree known as the *Local Fluff* or Local Interstellar Cloud (LIC) (see figure 4.7). This cloud measures a scant 10 light-years in diameter and amounts to a mere 0.3 solar masses. It can be directly sensed, as some of the interstellar material has leaked through the Sun-blown bubble of gas (the so-called heliosphere) and into the Solar System. In situ detections of extra-solar helium atoms and dust particles by interplanetary space probes have provided vital measures of this interstellar matter. In the next

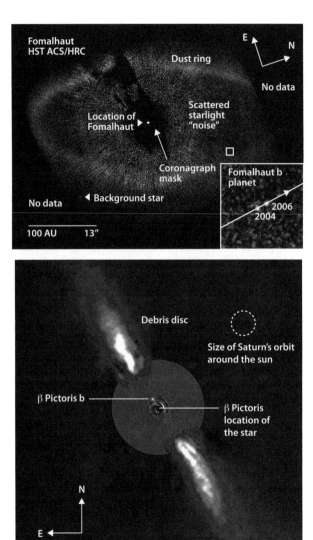

FIGURE 4.6. Recent discovery images of giant exoplanets around nearby stars. Light from the central stars has been blocked out to reveal much fainter features. (*Top*) Hubble Space Telescope images of the dusty ring of debris around Fomalhaut (Alpha Piscis Austrina) show a planet that appears to have cleared out a portion of the debris field. The planet is thought to have a bit less than three times the mass of Jupiter. (*Bottom*) Very Large Telescope composite image of the Beta Pictoris system shows a point-like source interior to the well-known debris disk (seen edge-on at near-infrared wavelengths). This source is estimated to be about eight times more massive than Jupiter. [(*Top*) P. Kalas, J. Graham, E. Chiang, and E. Kite (University of California, Berkeley), M. Clampin (NASA/Goddard Space Flight Center), M. Fitzgerald (Lawrence Livermore National Laboratory), K. Stapelfeldt and J. Krist (NASA/Jet Propulsion Laboratory), NASA, European Space Agency. (*Bottom*) A.-M. Lagrange (Laboratoire d'Astrophysique de Grenoble) et al., European Southern Observatory]

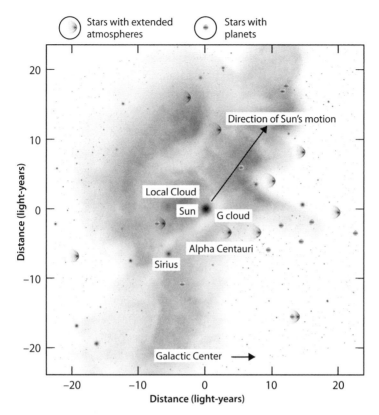

FIGURE 4.7. Schematic of our cosmic neighborhood, emphasizing the nebular components. The Sun appears to be moving through a diffuse cloud of gas and dust. The G cloud refers to the extended atmosphere associated with the Alpha Centauri system. [Interstellar Boundary Explorer (IBEX) Science Team, Wesleyan University, University of Chicago, Adler Planetarium, Goddard Space Flight Center, NASA]

10 years, the *Voyager 1* and *2* space probes will finally breach the heliopause—the surface that separates the heliosphere from the surrounding Local Interstellar Medium. These vintage spacecraft (launched in 1977) are still functioning, and so it is altogether likely that we will soon get our first distinct tastes of the Galaxy just beyond the Sun's domain.

The Local Fluff can also be observed through its absorbing effects on the light of stars in the solar neighborhood. High-resolution ultraviolet spectroscopy of these stars has revealed interstellar absorption in the lines of neutral hydrogen, neutral sodium, ionized calcium, and other ionized "metals." The cloud's inferred temperature of 6,300 Kelvins is hotter than the surface of the Sun but is considerably cooler than that of the even more tenuous bubble of gas that surrounds it. At a temperature of several *million* Kelvins, the so-called *Local Bubble* extends hundreds of light-years in all directions. Inside the evacuated bubble, hot gas can be observed as its five-times ionized oxygen absorbs ultraviolet light from more distant OB-type stars and white dwarfs. The coronal gas is also barely detectable in emission at X-ray energies. The walls of the Local Bubble and other adjacent bubbles betray their presence at radio wavelengths—producing vast loops of synchrotron emission in the sky.

Charting the Local Bubble is not for the faint of heart. There is very little in the way of "standard candles" to help one determine distances to the sundry sources of nebular emission or absorption. Most lines of sight manifest several spectroscopic "features," whose spatial relations are next to impossible to delineate. Those who dare to try, refer to the detailed spectra of stars along various lines of sight. Because the nearby stars have known distances, they can provide key milestones for the nebular features observed in the same direction. If one star shows no nebular absorption lines and the next farthest star does, then the source of nebular absorption must lie in between the two stars. Our current best guess of the Local Bubble's topology is shown in figure 4.8.

Recent extreme-ultraviolet observations with the *Cosmic Hot Interstellar Plasma Spectrometer (CHIPS)* have indicated that the Bubble's extent perpendicular to the galactic plane is significantly greater than that along the plane. If true, our Bubble may be venting its gas into the galactic halo. This is consistent with the violent origin of the Local Bubble itself. Astronomers estimate that inflating a bubble of such size and filling it with gas of such high temperature would require several supernova explosions over the last few million years. The source of this violence is thought to be the Scorpius-Centaurus association of young massive stars, some 500 light-years away. We find ourselves passing through the torrid aftermath of a relatively recent starburst. In time, the Local Bubble will cool to a fossil of its former self and slowly diffuse into the

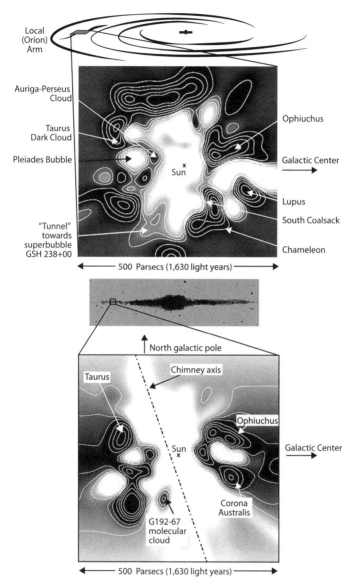

FIGURE 4.8. Cartoons of the Local Bubble as inferred from its absorbing effects on background stars. The nearest stars that show these effects delineate the outer edge of the Bubble. (*Top*) Face-on view of the Local Bubble. (*Bottom*) Edge-on view of the Local Bubble. [Adapted from press release at http://www.berkeley.edu/news/media/releases/2003/05/29_space.shtml, describing research by B. Y. Welsh (University of California, Berkeley), R. Lallement and S. Raimond (Université Versailles-St Quentin, Centre National de la Recherche Sciéntifique [CNRS], France), J.-L. Vergely (ACRI-ST, France), with reference to B. Y. Welsh, R. Lallement, J.-L. Vergely, and S. Raimond, "New 3D gas density maps of NaI and CaII interstellar absorption within 300 pc," *Astronomy & Astrophysics* 510 (2010): A54]

ambient interstellar medium. Other nearby hot stars will have gone supernova by then, inflating new bubbles of hot gas into the LISM. Meanwhile, we will have migrated out of this nebular hot-house and into some other percolating part of the galactic ecosystem.

Gould's Belt

Much of the galactic effervescence near the Sun can be attributed to Gould's Belt of star-forming activity. First noted in the 1800s by John Herschel during his stay in South Africa, Gould's Belt is most readily seen as a band of brilliant stars that courses around the entire sky. Benjamin Gould (producer of the massive *Cordoba Durchmusterung* star catalog and founder of the *Astronomical Journal*) was the first to trace the entirety of this feature from both northern and southern hemispheres, noting that a "great circle or zone of bright stars seems to gird the sky, intersecting with the Milky Way at the Southern Cross, and manifest at all seasons."

The band is inclined to the Milky Way by about 20 degrees, crossing the galactic plane near the constellations of Crux in the southern hemisphere and Cepheus to the north. It includes many of the well-known OB stellar associations, including the Scorpius-Centaurus association that is thought to have hosted the supernovae which created the Local Bubble (see the US-UCL-LCC complex in figure 4.9). Nearly half of all the bright stars in the sky can be attributed to Gould's Belt. These include the red supergiant Antares (Alpha Scorpii) along with its hot blue associates that delineate the constellation of Scorpius, the Garnet Star (μ Cephei)—one of the largest and brightest stars in the Galaxy, the green supergiant star Mirphak (Alpha Persei) which points the way to several nearby OB stellar associations in Perseus, as well as the blue supergiant star Rigel and the other brilliant blue stars that adorn Orion. As the astronomy author Ken Croswell wrote, "If I were kidnapped by an alien spaceship and taken to some remote corner of the Galaxy, Gould's Belt is what I'd look for to find my way back home."

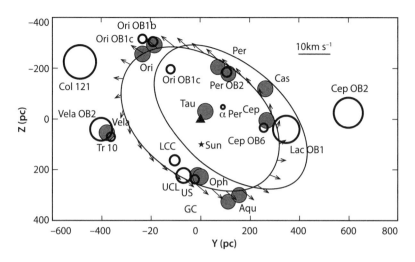

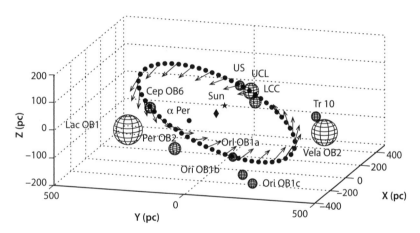

FIGURE 4.9. (*Top*) Gould's Belt of young OB stellar associations (empty circles) and star-forming clouds (shaded circles), where Z(pc) refers to the height above and below the galactic midplane. Characteristic velocities are shown with arrows. (*Bottom*) Perspective view of Gould's Belt showing only the OB stellar associations and velocities. In both renderings, the significant tilt of Gould's Belt with respect to the galactic midplane is evident. [Adapted from C. A. Perrot and I. A. Grenier, "3D dynamical evolution of the interstellar gas in the Gould Belt," *Astronomy & Astrophysics* 404 (2003): 519]

The estimated distances to stars in Gould's Belt indicate that it is not a perfect circle but rather an ellipse roughly 2,400 by 1,500 light-years in extent. The center of the ellipse is 500 light-years away in the direction of Taurus, roughly coincident with the Pleiades star cluster. The Sun is situated halfway between the center of the ellipse and the Scorpius-Centaurus OB association along the Belt's rim. Among the stellar luminaries in this celestial diadem, dusty clouds of atomic and molecular gas have been identified. Home to the next generation of stars, these clouds portend the future of our galactic neighborhood.

The gas clouds also provide vital evidence for the reality of Gould's Belt as an astrophysical phenomenon rather than a chance configuration of stellar groups. Spectroscopic analysis of the emitting nebulae confirm that the clouds and the stars of Gould's Belt are all moving as one coherently rotating and expanding system. The inferred expansion velocities (of order 10 km/s) combined with the overall size of the system suggest an expansion age of roughly 30–60 million years. This age is crudely coincident with the oldest stars associated with Gould's Belt. Further corroboration of orchestrated starburst activity comes from gamma-ray observations which have yielded an excess of compact sources along Gould's Belt. Such energetic phenomena can arise when supernova explosions leave behind super-active stellar remnants—likely in the form of rapidly rotating neutron stars. Given these sorts of happenstances associated with Gould's Belt, we are left speculating on the dynamical event(s) that gave rise to such an impressive creation.

Some astronomers suggest a spate of supernova explosions having detonated some 40 million years ago, and look to the OB stellar association centered on the Alpha Persei star cluster as the "smoking gun." Others contend that even this vociferous group of young stars could not have generated sufficient numbers of supernovae to generate the monolithic expansion and humongous pileup of gas that Gould's Belt represents. They also fret over the odd inclination of Gould's Belt to the galactic plane. Supernovae centered at some cluster would be more likely to foster pileups of gas within the galactic plane, where the ambient densities are greatest. For these reasons, other scientists have proposed that a massive gas cloud in the Milky Way's halo slammed into the galactic disk near the Sun. If the angle of impact was 20 degrees from the perpendicular, the collision could have generated an inclined blast wave that resulted in the expanding pileup of gas and stars which we recognize today as Gould's Belt.[6] Still others think that Gould's Belt represents a relatively nearby example of spiral-arm dynamics in the disk of our Galaxy—a mere blip of star-forming activity along one part of the so-called Orion Arm.

We will learn a lot more about the origin and evolution of Gould's Belt—as well as the rest of our galactic neighborhood—when the GAIA spacecraft carries out its mission to precisely reckon the distances and motions of more than a billion stars within 30,000 light-years of the Sun. Scheduled for launch sometime in 2013, this European mission will create the most accurate 3D map of the stellar Milky Way by far. It will also harvest a bounty of previously undetected planetary systems, thus enabling detailed investigations of planetary "demographics" in our sector of the Galaxy. Progress on this ambitious mission can be tracked at http://www.esa.int/science/gaia.

CHAPTER 5

•• • ● • ••

BEACONS FROM AFAR

Twinkle, twinkle, little star.
How I wonder what you are.
Up above the world so high,
Like a diamond in the sky!

—From "The Star" in *Rhymes for the Nursery*,
by Ann and Jane Taylor (1806CE)

AS WE LOOK BEYOND THE MEEK STARS of the solar neighborhood and past the dazzling upstarts in Gould's Belt, our view of stars in the galactic disk becomes increasingly compromised by the obscuring effects of interstellar dust. This irregular smokescreen puts a limit on how far we can see at visible wavelengths. In directions away from the murky Milky Way, our views become much clearer, allowing us to spy globular star clusters more than 50,000 light-years from us. This ability allowed Harlow Shapley to determine in 1920 that the system of globular clusters was centered well away from the Sun in the direction of Sagittarius. We have since confirmed that the overall architecture of the Milky Way is a halo-disk-bulge+bar affair with a common centroid at the radio source Sagittarius A*, some 27,000 light-years away.

Despite the limitations imposed by interstellar dust, there are several sightlines in the galactic disk that are relatively free of obscuration. These "windows" have allowed astronomers to perceive some structural features in the stellar distribution. The most luminous O- and B-type stars can be seen from the greatest distances and through the most dust. They also have the briefest lives—lasting only a few million years for the O stars to tens of mega-years for the B stars. For these reasons, the O and B stars provide ideal tracers of the recent star-forming activity and the structuring of this activity in the galactic disk.

Taken together, our various Earth-bound perspectives have yielded sufficient information for astronomers to decipher the types of stars that inhabit the Galaxy. We have also learned enough to perceive real differences among the stellar types with respect to their spatial distributions and motions. These differences offer vital clues to galactic astronomers intent on re-imagining the natural history of the Milky Way—a subject that we will more fully explore in chapter 11.

Motions and Distances

As "birds of a feather flock together," stars of common origin also share similar trajectories through space. This is most obvious for stars in clusters such as the Pleiades, Hyades, and Beehive (Praesepe). However, much looser arrangements of stars have been found by virtue of their common motions. These include the Sirius supercluster which we happen to be passing through, the Ursa Major moving group which contains many of the stars in the "Big Dipper," the Hyades supercluster which contains both the Hyades and Beehive (Praesepe) clusters, as well as the Scorpius-Centaurus OB association which has caused so much recent ruckus in the solar neighborhood.

To determine the space motion of a stellar group, astronomers measure each star's transverse ("proper") motion and its line-of-sight ("radial") velocity. The observed proper motion of a star across the sky is in angular units of arcseconds/year rather than linear units of velocity such as kilometers/second. Conversion to an actual velocity requires knowledge of the star's distance. For the nearest stars and corresponding groups, distances can be obtained from the measured heliocentric parallaxes. For the more remote stellar groups, the combined proper motions and radial velocities of the stellar members provide a handy way to determine this critical distance.

By mapping the proper motion "vectors" of each cluster member, and extending them forward, one can determine a convergent point on the sky—the celestial equivalent of the "vanishing point" in a perspective drawing. As shown in figure 5.1, the angle between this convergent point and the star corresponds to the angle that links the star's radial and transverse velocity vectors. The resulting trigonometric relations yield the desired distance for each member of the cluster and for the cluster as a whole.[1]

The textbook example of this distance gauging technique is the Hyades cluster. From radial velocity and proper motion measurements of some 200 stars in the cluster, astronomers have determined an average distance of about 46 parsecs (150 light-years). Recent measurements of the stellar parallaxes by the Hipparcos satellite give a more precise value of 46.3 parsecs. Reconciliation of these two measurements is of paramount importance, as the Hyades provides the fundamental stepping-stone for determining distances throughout the Galaxy and beyond to the rest of the universe. That is because the Hyades cluster is the closest stellar grouping that contains a sufficient variety of stars whose fundamental properties can be discerned. These stars, in turn, provide vital calibration standards for plumbing more distant stellar populations.

The Stellar Codex

From the combined use of heliocentric parallaxes and space motions to measure stellar distances, astronomers have assembled an impressive "Rosetta Stone" of fundamental stellar properties. The primary tool for visualizing these diverse properties is the Hertzsprung-Russell diagram, where some measure of luminosity is plotted against some index of temperature. As shown in plate 7 and figure 5.2, the resulting parameter space of stellar properties breaks up into discrete families. The most prominent family is the "main sequence" of stars which runs diagonally from the dim-red (lower-right) end to the bright-blue (upper-left) end of the H-R diagram. Comparison with theoretical lines of constant radius shows

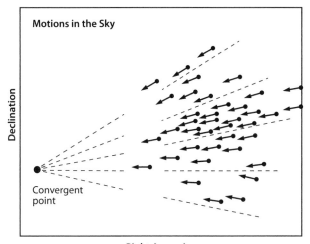

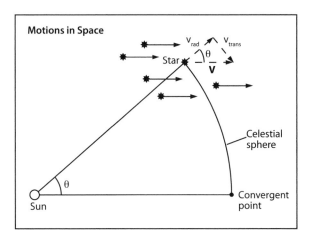

FIGURE 5.1. (*Top*) The "moving cluster" method for reckoning stellar distances relies on mapping the cross-wise "proper motions" of cluster members (in seconds of arc per year) and determining the convergent point of these proper motions on the sky. (*Bottom*) The angle between the convergent point and each star in the cluster is related trigonometrically to the transverse and radial velocities of that star. The transverse velocity, in turn, is related to the measured proper motion according to the star's distance (see endnote). [Adapted from *Introductory Astronomy & Astrophysics*, 4th edition by M. Zeilik and S. A. Gregory, Orlando, FL: Saunders College Publishing (1998), p. 384, with reference to *Elementary Astronomy* by O. Struve, B. Lynds, and H. Pillans, Oxford, UK: Oxford University Press (1959)]

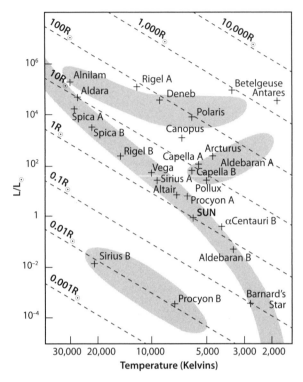

FIGURE 5.2. Hertzsprung-Russell diagram showing the surface temperatures and visual luminosities of well-known stars—several of which are naked-eye objects. Theoretical lines of constant radius are overlaid. These overlays are based on the assumption that stars radiate like "black bodies" in thermal equilibrium. The main-sequence stars range in luminosity from 1/1,000 solar to 100,000 times solar. Because their radii span a smaller range from 1/10 solar to 10 times solar, the tremendous variation in luminosity is mostly due to the corresponding variation in surface temperature. The giant and supergiant stars differ from the main-sequence stars, as they have much higher luminosities for their temperatures resulting from their much greater sizes. The white dwarfs are hot but dim, being much smaller than the main-sequence stars. [From Seeds, *Horizons: Exploring the Universe (with The Sky CD- ROM and InfoTrac)*, 7E. © 2002 Wadsworth, a part of Cengage Learning, Inc. Reproduced by permission. www.cengage.com/permissions]

that the main-sequence stars vary somewhat in size—but not enough to explain the fantastic range of luminosities. Instead, the wide-ranging luminosities can be attributed to the large variation in surface temperature that is evident along the main sequence.[2]

The family of giant stars is characterized by significantly higher luminosities at any given temperature. If these stars are radiating like thermalized "black bodies," then their much higher luminosities can be attributed to their much larger sizes. By the same reasoning, the white dwarfs are especially dim, because they are exceptionally small.

Stepping Out

Hertzsprung-Russell diagrams of individual star clusters provide further means for determining distances that take us well beyond the current range of parallax and proper motion measurements. Because all of the stars in a cluster are at approximately the same distance, their apparent brightnesses are mutually proportional to their absolute luminosities. We can take advantage of this proportionality to calibrate the observed color-magnitude diagrams (CMDs) of remote clusters with respect to well-calibrated H-R diagrams based on nearby stars (see plate 7), or with respect to theoretical models that are based on these empirical measurements.

Consider the handsome Pleiades star cluster (M45). From photometric observations of the individual stars, astronomers have obtained color-magnitude diagrams (CMDs), such as the one shown in figure 5.3. Here, the visual magnitude (m) is apparent, based solely on the measured flux of radiation through the yellow V-band filter. The cluster's CMD clearly shows a main sequence with apparent visual magnitudes ranging from V = 13 at the faint end to V = 3 for the brightest members. By fitting this main sequence to the "template" main sequence of absolute magnitudes and (B–V) colors, astronomers have found a difference between the apparent and absolute magnitudes amounting to (m–M) = 5.6 mags. This "distance modulus" corresponds to a distance of 132 parsecs (430 light-years), placing it beyond the solar neighborhood near the center of Gould's Belt. Through similar sorts of "main-sequence fitting," astronomers have determined the distances to many open star clusters within the galactic disk, and to the even more remote globular clusters that swarm throughout the galactic halo.[3]

Another important way to obtain stellar distances is via the "spectroscopic parallax" method. Through detailed analysis of a star's spectrum, it is possible to ascertain the star's spectral type

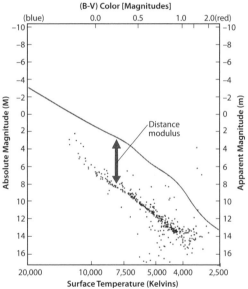

FIGURE 5.3. (*Top*) The Pleiades star cluster (M45) contains hundreds of stars. (*Bottom*) A color-magnitude diagram (CMD) of the stars making up the Pleiades shows a wide range of stellar colors and apparent magnitudes with a well-populated main sequence. By "fitting" the main sequence in this CMD with the main sequence on a fully calibrated Hertzsprung-Russell diagram (as represented by the sinuous line), astronomers can derive the distance modulus ($m - M$) and corresponding distance to this cluster. For the Pleiades, the measured distance modulus is 5.6 magnitudes, which yields a distance of 132 parsecs (475 light-years). [(*Top*) Courtesy of J. Foy (see http://picture.nsaac.org/gallery/main.php). (*Bottom*) Adapted from the interactive cluster-fitting program authored by K. Lee (University of Nebraska, Lincoln) which is available at http://astro.unl.edu/naap/distance/animations/cluster FittingExplorer.html]

and luminosity class to relatively high precision. Comparison with a similarly formatted H-R diagram (as in plate 7) will yield the star's absolute magnitude. Once the apparent magnitude is measured, it is a simple matter to compute the distance modulus (m–M) and corresponding distance. Pioneered by Jacobus Kapteyn in the early 1900s, spectroscopic parallaxes of clustered stars provide important checks on the distances that astronomers have obtained from main-sequence fitting. For single stars in the "field," the spectroscopic parallax may be the only game in town. That is, unless the star is an RR Lyrae or Cepheid variable.

These special cases are what Harlow Shapley used to fathom the Galaxy's system of globular clusters (see chapter 2). The supergiant Cepheids are sufficiently luminous to be detected in other nearby galaxies, as Edwin Hubble did in 1929—thereby determining that these nebulae were in fact autonomous "island universes." Once again, the trick is to find some proxy for absolute magnitude. In the case of RR Lyrae variables, their absolute visual magnitudes have been found to vary about a well-established average of about 0.6 mag. For the Cepheids, the "period-luminosity relation" discovered by Henrietta Leavitt in 1910 allows astronomers to monitor the star's variable light output, determine the period of variability, and then infer the star's absolute magnitude. Whenever these so-called standard candles are present, the job of fathoming distances becomes much easier.

Dancing with the Stars

So far, we have explored the astronomer's favored ways of reckoning stellar distances, luminosities, temperatures, and sizes. The most important stellar property remains to be determined, however. Governor of almost every stellar behavior and fate, the mass of a star is vexingly difficult to divine. Very little in the detailed spectrum of a star hints at the star's absolute mass. Indeed, a hot-blue main-sequence star can be just as massive as a red-cool supergiant star. We have slowly learned of these oddities by watching how stars dance with one another.

As Isaac Newton first proposed in 1687 and Albert Einstein elegantly explained 229 years later, mass and gravity are intimately coupled. Newton's Law of Universal Gravitation states the relationship most succinctly in terms of a gravitational "force" that acts between two masses. This invisible influence varies according to the product of the masses and the inverse of the square of their mutual separation, or

$$F = G \, m_1 m_2 \, / \, d^2$$

where F is the gravitational force, m is the mass, and d is the separation. The constant of universal gravitation G was first determined by Henry Cavendish in 1798 by carefully measuring the forces between suspended lead balls of precisely known weight. Armed with his far-reaching "inverse square" law, Newton could show that the trajectories of falling apples, whizzing cannonballs, and orbiting planets were all manifestations of the same essential dynamic.

Beginning with William Herschel's study of double stars, astronomers have capitalized on Newton's revolutionary relation to diagnose the masses of stars. All that was needed was to find stars under the influence of their mutual gravitation. Through careful telescopic examination of stars over time, astronomers had found myriad double stars that were in fact bound to one another in orbital fandangos. Each star in the binary system could be seen wheeling around some mutual center of mass that stayed put on the sky (see figure 5.4). By relating the gravitational force to the acceleration of the stars in their respective orbits, astronomers could relate the observed period of shared revolution to the combined masses and their mutual separation according to

$$P^2 \, (m_1 + m_2) = a^3$$

where the period P is measured in years, the masses m are in solar masses, and the separation a is in Astronomical Units. This relation is none other than Kepler's Third Law of Planetary Motion applied to binary stars. It is compromised by uncertainties in knowing the true separation, as the mutual orbits are likely inclined to our vantage by an unknown amount.

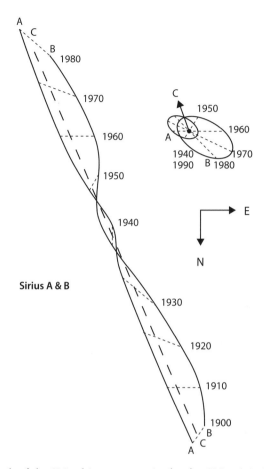

FIGURE 5.4. Path of the Sirius binary system in the sky. Sirius A is the bright A-type star that outshines all other stars in the naked-eye sky. Its much dimmer companion (Sirius B) is a white dwarf whose mass is roughly 1/2 that of the primary star. Each star revolves around the system's center of mass (C). [Adapted from *Introductory Astronomy & Astrophysics*, 4th edition by M. Zeilik and S. A. Gregory, Orlando, FL: Saunders College Publishing (1998), p. 237]

Another problem is that Kepler's Third Law only allows one to derive the summed masses of a binary star system. To ascertain the individual masses, it is necessary to determine the relative masses. This can be done by observing the relative distances between each mass and the mutual center of mass. As on a balanced seesaw, the body with the greatest mass will be located proportionately closer to the invisible pivot point according to

$$m_2 / m_1 = d_1 / d_2.$$

Although visual binaries such as the famous Sirius A+B system form the basis of all stellar mass determinations, most binary systems are either too far away or too tightly bound to be spatially resolved. Fortunately, their detailed spectra can be used to track the motions of the underlying stars. Through the Doppler effect on each star's pattern of spectral absorption lines, astronomers can monitor the blue- and red-shifting of spectral lines over time and thereby determine the system's orbital period as well as the ratio of masses according to

$$m_2 / m_1 = v_1 / v_2 = \Delta\lambda_1 / \Delta\lambda_2,$$

where the relative velocities (v_1 / v_2) can be directly derived from the relative Doppler shifts in wavelength $(\Delta\lambda_1 / \Delta\lambda_2)$. This dynamical method for weighing stars turns out to be very difficult, with many details and challenges needing attention. Despite more than 200 years of valiant effort, astronomers have measured the masses of only a couple hundred stars. Most of these determinations come from eclipsing binary stars with each of the spectral patterns resolved. Here, the orbital inclinations are well known to be perfectly edge-on, thus allowing reliable determinations of the individual masses. What they have found can be depicted in one plot of vital importance (see figure 5.5). The mass of a main-sequence star is seen to completely determine its luminosity according to

$$L / L_{\text{Sun}} = (m / m_{\text{Sun}})^n.$$

For intermediate-to-high stellar masses (0.5 solar masses to 20 solar masses), the exponent $n \cong 4.0$. This is a far cry from a linear relation, where a doubling of mass yields a doubling of luminosity. Instead, the luminosity is seen to skyrocket by more than an order of magnitude for every doubling of mass. Surely, the observed mass-luminosity relation must have consequences over time. How long can high-mass stars maintain their profligate ways?

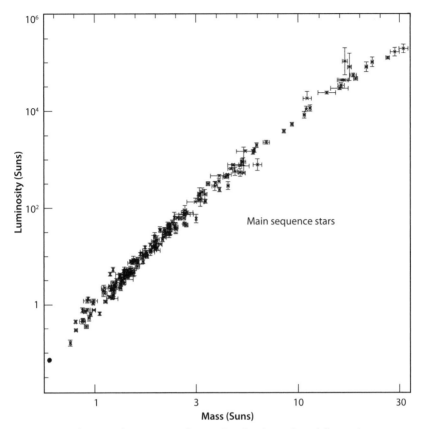

FIGURE 5.5. The mass-luminosity relation that has been found for main-sequence stars shows a steep dependence, with small increases in mass yielding huge increases in luminosity. [Courtesy of O. Y. Malkov, with reference to O. Y. Malkov 2007, "Mass-Luminosity Relation of Intermediate-mass Stars," *Monthly Notices of the Royal Astronomical Society* 382 (2007): 1073–1086]

A crude idea of the total stellar lifetime T can be obtained by dividing the available "fuel" m by the vigor of the "fire" L. More specifically,

$$T / T(\text{Sun}) = (m / m_{\text{Sun}}) / (L/L_{\text{Sun}}),$$

which reduces to

$$T / T(\text{Sun}) \cong (m / m_{\text{Sun}})^{-3.0}.$$

A good estimate for the Sun's total thermonuclear lifetime is roughly 10 billion years. Therefore, the lifetimes of other stars go as $T \cong 10^{10} \, (m / m_{Sun})^{-3.0}$ years, a reasonable approximation to what the more sophisticated stellar models indicate. For example, a 10 solar-mass B3 main-sequence star has a predicted lifetime of only 10 million years, while a 0.7 solar-mass K5 star should last about 30 *billion* years—much longer than the 13.7 billion-year age of the universe itself!

Galactic Archaeology

Our consideration of binary stars has taken us a long way. We are now fully equipped to understand the Milky Way's diverse stellar populations in evolutionary terms. One way is to simply regard a Hertzsprung-Russell diagram, where lines of constant lifetime (isochrones) are plotted (see figure 5.6). Here, we see that the upper main-sequence and supergiant stars are typically much younger than the lower main-sequence and red-giant stars.

Another way is to compare the H-R diagrams of individual star clusters (see figure 5.7). Besides being at the same distance, the stars in a particular cluster all formed at about the same time. Therefore, clusters of differing age will have main sequences that are populated to differing degrees. The H-R diagram of the young double cluster h + chi Persei shows this effect, with the main sequence fully populated at the high-luminosity end. By contrast, the Praesepe (Beehive) cluster in figure 5.7 is missing its high-luminosity stars—a sure sign that the cluster has been around for a while. An examination of the globular clusters in the halo of the Galaxy reveals even more variation. In many cases the main sequence is barely evident as a low-luminosity stump, with many of the stars in giant phases consistent with their lower masses (see figure 5.8). By reckoning the extent of the main sequence and the situation of the giant phases, astrophysicists have come to the conclusion that most of the Milky Way's globular clusters are incredibly old—with ages ranging from 11–14 billion years. We find ourselves witness to surviving relics of the early universe—when the Milky Way was in its infancy.

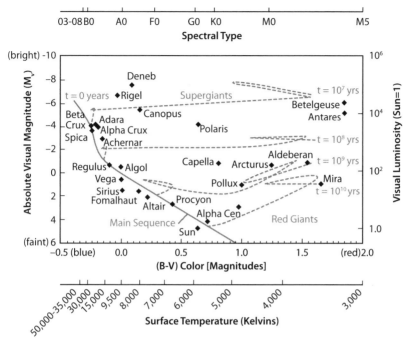

FIGURE 5.6. Hertzsprung-Russell diagram showing the brightest naked-eye stars along with modeled lines of constant stellar lifetime. The high-luminosity part of the diagram is dominated by stars that do not last long and hence are of recent birth. Good examples of these short-lived wonders are Deneb and Betelgeuse. The low-luminosity part is dominated by stars that can last a very long time, and hence are likely to be very old. The red giants Aldebaran and Mira are good examples of this hoary genre. [Adapted from *Galaxies and the Cosmic Frontier* by W. H. Waller and P. W. Hodge, Cambridge, MA: Harvard University Press (2003), p. 44]

Denizens of the Disk

Beginning with Kapteyn's seminal surveys in the late 1800s, astronomers have endeavored to create complete dossiers of stellar distances, masses, ages, and motions. Although the obscuring effects of dust and the confusing consequences of crowding in the galactic disk are especially troublesome, astronomers have made considerable progress by scrutinizing the stars within 5,000 light-years of the Sun. This piece of galactic real estate is sufficiently large for general trends and patterns to be identified.

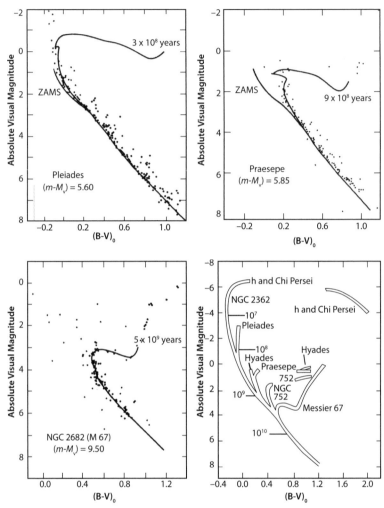

FIGURE 5.7. H-R diagrams of open (or "galactic") star clusters in the disk of the Milky Way reveal clear differences in the populations of their main sequence and giant branches. These differences can be used to determine the ages of the respective clusters. The clusters and ages represented here are the open clusters M45 (Pleiades), M44 (Praesepe), and M67. ZAMS refers to the Zero-Age Main Sequence. (*Bottom right*) A composite H-R diagram of many open star clusters shows that the youngest star clusters fully populate the Main Sequence to high luminosity, while the older clusters are lacking in high-luminosity Main Sequence and Giant stars. The ages along the Main Sequence refer to the point on the diagram where a cluster's stars of that age "turn off" the Main Sequence. [(*Top left, top right, bottom left*) Adapted from *Introductory Astronomy & Astrophysics*, 4th edition by M. Zeilik and S. A. Gregory, Orlando, FL: Saunders College Publishing (1998), p. 327. (*Bottom right*) Adapted from *The Milky Way*, 5th edition by B. J. Bok and P. F. Bok, Cambridge, MA: Harvard University Press (1981), p. 119 with reference to H. L. Johnson and A. R. Sandage, *Astrophysical Journal* 121 (1955): 616–627]

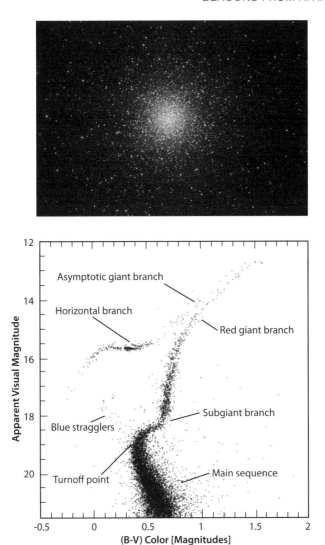

FIGURE 5.8. The globular cluster M3 and its H-R diagram. The cluster is centrally concentrated and symmetric, befitting its ancient age and dynamically "relaxed" state. Its H-R diagram is completely lacking in high-luminosity massive stars, as they died off billions of years ago. Instead, the plotted population consists exclusively of long-lived low-mass main-sequence stars along with giant stars which will soon finish their life cycles. [(*Top*) E. Gellman, D. Creighton, and Adam Block, Advanced Observers Program, National Optical Astronomy Observatory (NOAO), Associated Universities for Research in Astronomy (AURA), National Science Foundation (NSF). (*Bottom*) From *Galactic Astronomy* by J. Binney and M. Merrifield, Princeton, NJ: Princeton University Press (1998), p. 334]

Most telling are the variations in spatial distributions and motions among stars of different types. The O- and B-type stars, for example, are found in nearly circular orbits very close to the galactic midplane. These short-lived tracers of recent star-forming activity inform us that most of the stellar birthing in the Galaxy is currently confined to a very thin layer no more than 400 light-years in vertical extent. The distributions and motions of supergiant stars, molecular clouds, emission nebulae, and other tracers of recent star-birth activity share these traits—further corroborating the idea that the current birthplaces of stars are for the most part restricted to a very thin disk.

By contrast, stars of lower luminosity, lower mass, longer lifetimes, and characteristically older birth dates can be found strewn ever farther from the galactic midplane. Moreover, their motions are decidedly less circular and co-planar, with significant radial and vertical components inferred from their measured space velocities (see figure 5.9).

A good case in point is the Sun itself. This G-type main-sequence star is thought to have a total thermonuclear-powered lifetime of about 10 billion years. It is currently halfway through its hydrogen-fusing main-sequence phase, with another 5 billion years to go before it exhausts the hydrogen fuel in its core, ignites H-fusion in a shell, and bloats into a red giant. Consistent with its age, the Sun strays significantly from pure circular co-planar motion.

By tracking the trajectories of stars near the Sun, astronomers have inferred that a good portion of these motions are in fact reflections of the Sun's own peculiar motion.

Situated twenty-five light-years above the galactic midplane, the Sun is traveling in the direction of the constellation Hercules at a mean speed of 13.4 km/s—about half the Earth's orbital speed around the Sun. The Sun's peculiar motion in the disk includes a vertical component of 7.2 km/s, a radial (inward trending) component of 10.0 km/s, and a tangential motion that leads the average galactic rotation near the Sun by 5.2 km/s. This amounts to little more than 4 percent of the solar neighborhood's overall rotational speed of 220 km/s. The Sun's vertical velocity component causes it to bob up and down through the galactic midplane with a period

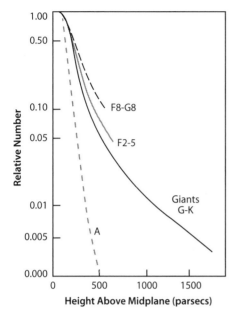

FIGURE 5.9. Distributions of stars above and below the galactic midplane according to their spectral types. The A-type main-sequence stars hew much closer to the galactic midplane compared to the typically older F and G main-sequence stars and G and K giant stars. [Adapted from *Galactic Astronomy*, 2nd edition by D. Mihalas and J. Binney, New York: W. H. Freeman (1981), p. 250]

of about 42 million years. Its radial and tangential motions also oscillate, yielding a net elliptical orbit (4 percent deviation from circular) which is centered on the galactic nucleus and which slowly precesses around the Galaxy every 170 million years. The combined motion of the orbiting Sun produces a rosette pattern in the galactic plane akin to the designs you can make with a spirograph. All of this excess motion is consistent with the peculiar motions of other middle-aged G-type stars like the Sun.

How did the stars differentiate themselves this way? What succession of events over time could have led to such diverse configurations and motions?

One scenario posits that the juvenile Milky Way sported a much thicker disk of unsettled gas, in which the earlier stars formed. Since then, the gas has settled down into a thin disk with all deviant motions collisionally dissipating away and only circular co-planar orbits remaining. This scenario can explain the age-dependent trends in stellar distributions and motions. It is especially successful at explaining the so-called thick disk of stars that has been identified extending more than 3,000 light-years from the galactic plane and rotating at a grandmotherly speed of only 180 km/s. However, it challenges astronomers to devise some means of forming stars from a fatter disk of more diffuse gas.

The alternative is to imagine the gaseous disk having gravitationally and collisionally settled down over a short period of time following the Milky Way's birth. All stars have since formed within this thin disk, but over time, the older stars have suffered occasional perturbations that jostled them into increasingly disparate orbits. Prime agents of perturbation include the itinerant star clusters and molecular clouds that a star would ultimately encounter during its multiple revolutions around the Galaxy. Like the juvenile thick-disk scenario, this persistent thin-disk model also explains the age-dependent stellar wanderings, but with the added cachet of preserving the dense gaseous circumstances of star formation as we observe them today. To explain the even more extended thick disk, many astronomers today invoke outside influences in the form of a dwarf galaxy that merged with the Milky Way some 10 billion years ago and has since diffused throughout the disk (see chapter 11). Perhaps in the next decade or two, we can test these two evolutionary scenarios by mapping the distributions and motions of newborn stars within the "young" spiral galaxies that we have identified at great distances and corresponding "lookback times." If the primordial disks turn out to be thin and in circular rotation, then we will know that spiral galaxies like the Milky Way settled down very quickly. That determination will require considerably greater acuity and sensitivity than we have today, however.

Besides manifesting vertical differentiation, the differing stellar types also show variegated spatial arrangements as imagined from above the galactic plane (see figure 5.10). From this idealized bird's-eye perspective, the clusters and associations of luminous young O- and B-type stars appear clumped into curved bands. Believe it or not, these poorly defined stellar arcs represent some of our best evidence for spiral structure in the Milky Way Galaxy!

Unlike the hot young stellar upstarts, the lower-luminosity red giants and main-sequence stars do not share this patterning. Instead, these ruddy stars appear uniformly spread throughout the

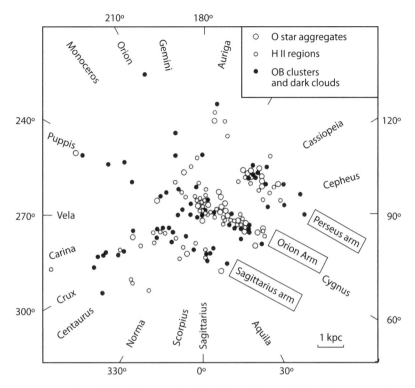

FIGURE 5.10. Spatial distribution of HII regions, clustered O and B stars, and associated dark clouds within 10,000 light-years (3,000 parsecs) of the Sun, as viewed from above the galactic plane. These various tracers of recent star formation appear to demarcate fragments of three spiral arms. A more comprehensive (but more confused) rendering of this same region can be accessed at http://www.atlasoftheuniverse.com/nebclust.html. [Adapted from *The Milky Way*, 5th edition by B. J. Bok and P. F. Bok, Cambridge, MA: Harvard University Press (1981), p. 272, based on data from W. Becker and Th. Schmidt-Kaler]

portion of the disk that is accessible to our visible reckoning. The elder stars were most likely created in spiral arms, but have since circuited the Galaxy many times, and in doing so, meandered away from their original configurations.

The Milky Way's stellar disk has a few more tricks to challenge and delight astronomers. In the direction of Cygnus, the run of stars rises above the nominal midplane. Opposite to this direction—toward Vela—there is an excess of stars below the midplane. Astronomers suspect that these deviations are tracing a palpable

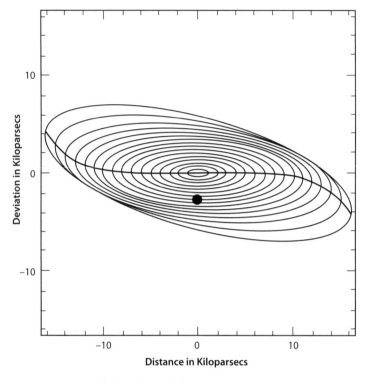

FIGURE 5.11. Warping of the galactic disk, as mapped by the 2MASS near-infrared imaging survey. In this depiction, the Sun's location is indicated by the black dot. [Adapted from M. López-Corredoira et al., "Old Stellar Galactic Disc in Near-plane Regions According to 2MASS: Scales, Cut-off, Flare and Warp," *Astronomy and Astrophysics* 394 (2002): 883–899]

warping of the stellar disk. The optical Hipparcos mission and near-infrared 2MASS imaging survey have since mapped this warping, showing that the galactic disk has been twisted like the brim of a fedora (see figure 5.11). Similar warping is evident in the more extensive distribution of atomic hydrogen. The twisting agent remains controversial, but the Large and Small Magellanic Clouds remain prime suspects due to their substantial masses and close proximity. In one intriguing scenario, the passage of the Magellanic Clouds through the Milky Way's dark-matter halo is thought to have produced a disturbance in the halo akin to the "wake" that a motorboat leaves upon the water. The disk of the Milky Way has since responded to the gravitational effects of the reconfigured halo, thus explaining the observed warping.

Ghosts in the Galactosphere

You do not have to look very far to find stars belonging to the halo, as several of these itinerant balls of fiery plasma happen to be passing through our sector of the galactic disk. We can recognize them based on their unusual motions and their distinctive spectra. From their radial velocities, we see that they do not partake in the rotation that is common to the stars of the nearby disk. Instead, they seem to be approaching us or receding from us at drastic speeds. Once called "high-velocity stars," they are now recognized to be just the opposite. The Sun and other disk stars are, in fact, revolving fastest around the Galaxy. From our whizzing perspective, however, the laggard halo stars in the nearby disk are perceived to have unusually high velocities of approach or recession.

The nearby halo stars also manifest spectra that are unusually bereft of absorption features associated with iron, magnesium, calcium, and other "metals." The effect is especially pronounced in the far-violet part of the spectrum, where many metal absorption lines can be found. The net result is a "purer" continuous spectrum with a lot more radiant power in the violet. When plotted on an H-R diagram, stars of this ilk will have bluer colors for their overall luminosity—what is known as an "ultraviolet excess." Astronomers first interpreted their placement on the H-R diagram as indicating lower luminosities for their spectral types, and so dubbed them "subdwarfs." This somewhat misleading name has stuck—with the subdwarf category today being identified with metal-poor high-velocity stars that belong to the ancient halo or to the transitional thick disk. Thanks to the nearby subdwarfs, we can get a taste for the swarming ways that characterized the primordial Milky Way.

Although much of the halo is occupied by single subdwarfs and their evolved red-giant counterparts, the most prominent players in the halo are the globular clusters. Host to tens of thousands to upwards of a million stars, each globular cluster looks like a bejeweled broach on a black velvet background (see plate 8). A typical globular cluster's H-R diagram (shown in figure 5.8) is characterized by a severely truncated main sequence that is shifted blueward according to its "ultraviolet excess." There are no stars occupying the supergiant branch of the H-R diagram, but the lower-luminosity red-giant branch, horizontal branch, and asymptotic giant branch are fully populated. Here we are witness to the final evolutionary stages of Sun-like stars that have already exhausted the hydrogen in their cores. By carefully measuring the main-sequence "turnoff" and other telling features in the H-R diagram, astronomers have determined that most globulars are downright decrepit, with characteristic ages of 11–14 billion years. One can only imagine what they were like when they still had their most massive and luminous stars.

Indeed, it takes a whole lot of imagining for one to explain how the globulars came to populate the nearly vacant halo in the first place. Did they form out of the proto-galactic nebula before it settled down into a disk? Or did they form inside dwarf galaxies that were then captured by the larger Milky Way? Although we are far from answering these questions in any definitive way, we have found some tantalizing clues in the overall distribution of stars across the sky.

Like jet contrails in an azure sky, myriad streams of stars can be tracked across the *Galactosphere*. These first turned up in all-sky surveys on photographic plates. More sensitive digital surveys at optical and infrared wavelengths have since confirmed the presence of several star streams coursing through the galactic halo (see plate 8). In some instances, the streams can be traced back to globular clusters or dwarf galaxies within the Milky Way's domain of tidal influence. For example, optical radial-velocity studies of stellar fields near the galactic bulge revealed in 1994 a co-moving group of stars that does not share the general motions in this part of the Milky Way. The optical Sloan Digital Sky Survey, near-infrared 2MASS sky survey, and other subsequent surveys have confirmed the existence of a separate diffuse dwarf galaxy in close proximity to our own galaxy. Dubbed the Sagittarius Dwarf Elliptical Galaxy (SagDEG), it is likely in the process of being ingested by the Milky Way Galaxy. As it is ripped apart by the Milky Way's intense gravity, this hapless galaxy is leaving behind a tidal tail of red giants that has been observed to extend around more than 360 degrees of the sky (see figure 5.12).

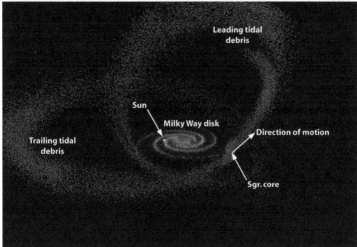

FIGURE 5.12. (*Top*) The Sagittarius Dwarf Elliptical Galaxy (SagDEG) presents a barely detectable enhancement of star counts in the direction of Sagittarius. It is located on the far side of the Milky Way, at an estimated distance of 50,000 light-years. (*Bottom*) Depiction of the stellar stream associated with the Sagittarius Dwarf Elliptical Galaxy (SagDEG). The Sun's location is shown as a dot. The densest part of the stream represents the dwarf galaxy's core. Tidal debris from the stripped galaxy can be seen extending from the core, wrapping around the galaxy, and descending near the Sun's position. [(*Top*) Adapted from R. Ibata (University of British Columbia), R. Wyse (Johns Hopkins University), and R. Sword (Institute of Astronomy), with reference to R. Ibata et al., "The Kinematics, Orbit, and Survival of the Sagittarius Dwarf Spheroidal Galaxy," *Astronomical Journal* 113 (1997): 634–655. (*Bottom*) Courtesy David Law, University of Virginia. (See http://www.astro.ucla.edu/~drlaw/Sgr/)]

About a dozen stellar streams are currently known, though there may be several hundred ghostly trails of shredded galaxies and globular clusters still fossilized in the halo. The presence of these relics lends credence to the notion that a good bit of the Milky Way was built up from the merging of many smaller galaxies. Perhaps most, if not all, of the globular clusters are the hardy remnants of these galaxy mergers. Of course, the devil is in the details, with the history of hierarchical galaxy construction still more of an art than a science. Most current theories of galaxy formation rely on cold dark matter haloes to get things started. The dark haloes attracted the ordinary (baryonic) matter, which then spawned the stellar systems we today call galaxies. These theories predict a lot more dark-matter-dominated dwarf galaxies buzzing around the Milky Way than are currently seen. Perhaps we will soon find evidence for the erstwhile satellite dwarfs, as we obtain ever more sensitive mappings of the stellar lacework in the galactic halo.

· · ● ● · ·

CHAPTER 6

STAR BIRTH

How is it that the sky feeds the stars?
—From *On the Nature of Things* by Titus Lucretius Caras (55 BCE)

Regions of lucid matter taking forms,
Brushes of fire, hazy gleams,
Clusters and beds of worlds, and bee-like swarms
Of suns, and starry streams.
—From "The Palace of Art" in *Poems by Alfred Tennyson* (1833 CE)

TO AN ASTRONOMER, the birth of a star is as miraculous as the birth of a child. Both creative processes begin with essentially nothing and culminate in fully formed wonders—as exquisite as they are expressive. Of course, there are important differences between spawning baby humans and infant stars. For starters, the gestation of a human begins with a tiny fertilized egg that then draws on the mother's placenta for all of its material and energetic needs. Stars do not form from such minuscule seeds that subsequently subdivide, feed on their surroundings, and grow into animate complexes populated by more than 10 trillion metabolizing cells. Instead, stars are *distilled* from vast diffuse clouds of gas and dust that somehow find ways to concentrate their holdings by incredible amounts.

In Darkness Born

Consider a typical stellar nursery—shepherded into coherence by the incessant churning of interstellar turbulence, or by some passing spiral density wave, supernova shock wave, or collection of windy stars. The resulting cloud has spatial dimensions of tens to hundreds of light-years and contains tens to hundreds of *thousands* of solar masses—mostly in the form of cold molecular hydrogen. By any terrestrial measure, the cloud is extraordinarily tenuous, with densities of a few hundred molecules per cubic centimeter. This is less than any vacuum we can create in the laboratory—the equivalent of exhaling a single lungful of air molecules into a volume the size of the Alps! The distillation of a star from such a cloud requires contracting spatial dimensions by an even greater factor of about ten million (10^7). The corresponding material condensation amounts to a factor of $(10^7)^3$, or a billion trillion (10^{21}) in density. This is tantamount to taking that same lung-exhaled mountain range of highly diffused air molecules and contracting it to a droplet of water. Wow!

The closest example of such a stellar nursery is the Taurus Molecular Cloud. Situated 450 light-years from us, this filamentary cloud can be seen in silhouette against the background of faint stars in the constellation of Taurus the Bull (see figure 6.1). Careful measurement of the dusty cloud's obscuring effect on background stars has yielded precise maps of the cloud's material contents. Millimeter-wave spectral-line surveys in the light of carbon monoxide (CO) have produced similar renderings. These complementary mappings of the Taurus Molecular Cloud reveal a cloud of

FIGURE 6.1. Visualizations of the Taurus Molecular Cloud. (*Top*) Historic photograph taken by E. E. Barnard in 1875 showing nebulous regions of darkness contrasted against the ambient star field. (*Middle*) Image of the visual extinction in the direction of the Taurus Molecular Cloud, as derived from near-infrared photometry of the background stars and the inferred degree of stellar reddening. The darker regions indicate greater amounts of stellar reddening and obscuring dust along these lines of sight. The rectangle delineates the field in E. E. Barnard's photograph. (*Bottom*) The complementary distribution of mm-wave emission from carbon monoxide molecules in the Taurus Molecular Cloud. This image was crafted by combining data from approximately 3 million spectra obtained with the Five College Radio Astronomy Observatory 14-m diameter radio telescope in central Massachusetts. The displayed range of brightness scales with the total amount of carbon monoxide gas along the line of sight. [(*Top*) From *A Photographic Atlas of Selected Regions of the Milky Way* by Ed-

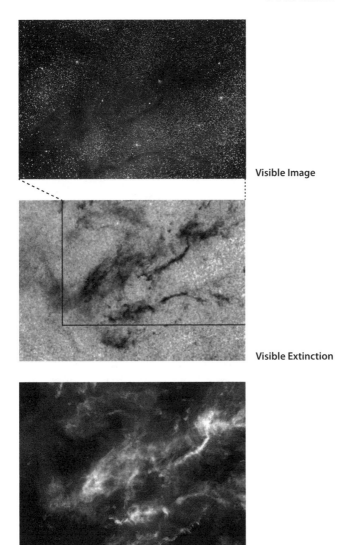

Visible Image

Visible Extinction

CO Emission

ward Emerson Barnard, ed. Edwin B. Frost and Mary R. Calvert, Washington, DC: Carnegie Institution of Washington (1927). Atlas plate 5 accessed from the Georgia Tech digital collections on E. E. Barnard at http://www.library.gatech. edu/search/digital_collections/barnard/index.html. (*Middle, Bottom*) Courtesy of Paul Goldsmith (Jet Propulsion Laboratory, Caltech, NASA), with reference to J. Pineda, P. F. Goldsmith, N. Chapman, et al., "The Relation of Gas and Dust in the Taurus Molecular Cloud," *Astrophysical Journal* 721 (2010): 686] modest size (90 by 70 light-years), intermediate mass (24 thousand solar masses), and variable density with a mean of 300 molecules per cubic centimeter. The cloud's temperature is incredibly low—just a few degrees Kelvin above absolute zero! More of the cloud's kinetic energy can be found in turbulent motions among the sundry cloud parcels. This turbulence is "supersonic," though it amounts to only a few tens of km/s. By and large, we are looking at a very cold, dark, and quiescent nest for breeding stars.

Nestled within Taurus, about a dozen dense cloud cores have been detected in the microwave glow of CS, $C^{18}O$, N_2H^+, and other molecules that are well-suited to tracing these gaseous knots. Here, typical scales are a few tenths of a light-year, with masses amounting to a solar mass or so, and densities of a few thousand up to about a million molecules per cubic centimeter. Many of these cores are arrayed along narrow filaments of gas which help to define the overall structure of the larger cloud. The cores and hosting filaments contain most of the young stellar objects that have been found in Taurus. Infrared observations are most favored to picking up the telltale signs of the newborn stars, as these embryonic sources are often still swaddled in blankets of dust which absorb the starlight, heat up, and re-radiate the star's energy in the infrared.

The more diffuse gas appears to have been combed by some enormous brush. Many astronomers look to large-scale magnetic fields as a vital structuring agent. As the fields thread through the cloud, they cause the gas to align with its sinuous lines of force. Perhaps the magnetic fields play an important role in mediating the condensation of the gaseous cores, and of the young stellar objects within. To learn more, we need to look closer.

Stages of Stellar Gestation

Although astronomers are hard-pressed to follow the complete star-forming process, they have learned enough to identify several key stages (see figure 6.2). These include:

Cloud Aggregation, where the molecular cloud is somehow put together. Various sweeping activities can do the trick. In far-infrared maps, we see evidence for stellar winds or supernova blasts having plowed up the interstellar medium into sundry shells and filaments (see figure 3.7). Spiral density waves could also play a role, as small cloudlets crowd together during their passage through the interstellar traffic jams that the density waves induce. The pell-mell turbulence that characterizes the interstellar medium is also likely to play a role. We can see the signature of turbulence in the hierarchical structuring of nebular material. Like other "fractal" forms in nature (snowflakes, trees, coastlines, etc.), small clouds are clustered into larger clouds which, in turn, are organized into even larger constructs of similar relative spacing and sizes. This "self-similar" architecture naturally arises as turbulent energy cascades from larger to smaller scales.

Once the cloud is cobbled together, collisions between the gas particles in the cloud ramp up. The energy of the collisions causes the atoms and molecules in the gas to attain higher levels of quantized excitation. This lasts only a few nanoseconds, as the excited particles then radiate away their excess energy. The result is an even cooler and denser cloud. Meanwhile, the dust grains in the cloud provide enhanced protection from the disruptive ultraviolet photons that are spewing from any nearby hot stars.

Core Condensation, where gravity begins to take hold and fragment the cloud into several much denser cores. The threshold for gravitational condensation occurs when the parcel's gravitation exceeds all resistive agents. These include the parcel's thermal, turbulent, and magnetic energies. In the 1920s, Sir James Jeans articulated the energetic prerequisites for gravitational fragmentation and condensation. Consequently, we call the threshold mass for gravitational instability the *Jeans mass*—above which gravitational condensation is assured. As you might expect, this threshold increases with the (resisting) temperature of the cloud and decreases with the (abetting) cloud density. For a cloud like Taurus, typical Jeans masses amount to several solar masses. No wonder Taurus is a favored locale for spawning intermediate-mass stars like the Sun along with lower-mass stars and brown dwarfs.

Gravitational Collapse, where the core gravitationally implodes. Once gravity gets the upper hand, there is pretty much nothing

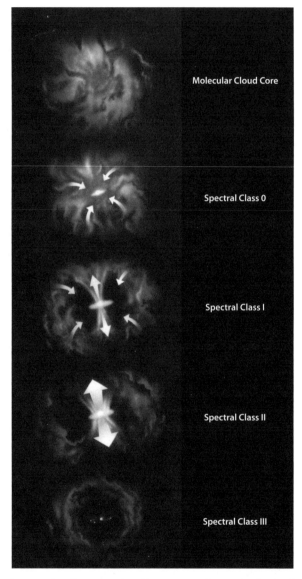

FIGURE 6.2. Stages of stellar gestation (*from top to bottom*)—beginning with a molecular cloud core, the infall of gas to form a Young Stellar Object (YSO) with a growing protostar and accretion disk (Spectral Class 0), the development of a strong bipolar outflow that emanates from an accreting YSO (Spectral Class I), the clearing of material from the YSO's disk and envelope (Spectral Class II), and a fully visible pre-main-sequence star with surrounding planets and debris (Spectral Class III). [Courtesy of Charles Lada (Harvard-Smithsonian Center for Astrophysics) and Rob Wood (Illustrator)]

that can halt the collapse. This occurs essentially in free-fall from the inside out. The time-scale for free-fall collapse depends only on the core's density—the higher the density, the quicker the collapse. Typical molecular cloud cores have densities of up to a million molecules per cubic centimeter. Their corresponding collapse times are about a million or so years. That would suggest very rapid conversions of molecular cloud cores into protostars. However, we can observe molecular clouds chock-a-block with dense cores throughout the Galaxy—as if some sort of "galactic prophylactic" is preventing the wholesale spawning of stars in an orgy of collapsing cores. Somehow, most cores are resistant to collapse, with only a few on the brink at any particular time. Likely inhibitors include the core's rotation and internal turbulence.

For those cores in free-fall, the Jeans scenario for fragmenting, condensing, and further fragmentation would continue forever, until all that remained was a flurry of protoplanets. That outcome does not happen, however, as new agents of resistance kick in to halt any further breakups. These agents are provided by the core itself. Previously transparent to its own infrared radiation, the core eventually gets sufficiently dense to interact with the emergent photons. In so doing, it begins to heat up. The resulting thermal energy contained within the warm core eventually halts the free-fall, constraining any further contraction to a rate that is regulated by the core's luminosity.

Mediated Contraction, where the gravitational energy of the infalling material is converted into thermal energy of the protostellar core. At this point, we need to consider the actual mechanisms that mediate the contraction of material onto the incipient protostar. Angular momentum is a critical arbiter of the contraction, because any initial rotation of the core is amplified by further contraction, until the core has spun up to its gravitational limit. To surpass this limit, the core needs to shed some of its angular momentum. Astronomers pretty much agree that the best way to do this is to remove or transfer some of the rotating mass away from the protostellar core. This can be done by breaking up the core into a binary system of two smaller cores in rotation about their mutual center of mass. Another way to remove angular momentum from the shrinking core is to build up a disk of rotating material that surrounds the growing protostar. The rotation of the disk provides a handy reservoir for storing the excess angular momentum. It also provides an important means for transporting material from the core's outer envelope to the growing protostar. These sorts of circumstellar disks are known as *accretion disks*. A third way to shed rotating mass is via the emission of bipolar jets along the core's rotational axis. This mechanism can work just as long as the outflowing jets also have considerable rotational motion. We will meet marvelous examples of the jet phenomenon later in this chapter.

The contracting core also needs to get rid of its magnetic field. As the core contracts, whatever magnetic field that threaded through it intensifies until it becomes a resistive agent against further collapse. The magnetic field acts only upon the minority of gas particles that are charged (the ions), but that is enough to retard the overall contraction. Astrophysicists have wrestled with this issue and have come up with a mechanism called *ambipolar diffusion*, which describes how the charged and neutral particles interchange energies in such a way for the neutrals to slowly drift inward past the magnetically fixed ions.

The final picture is of a core that is collapsing preferentially in directions mediated by its overall rotation and magnetic field, with new material accreting equatorially, and excess material squirting out along the rotational axis. Whew! Meanwhile, the *rate* of contraction is carefully regulated according to the core's ability to radiate away its gravitational energy. The outcome is a *protostar* that is being powered by gravitational contraction alone. At first, the protostar is completely cloaked in residual gas and dust. Its spectrum is dominated by the infrared light from dust that is re-radiating the luminosity from within. Astronomers call such infrared-dominated protostellar sources Young Stellar Objects, or YSOs, of spectral Class 0 (see figure 6.2).

Competitive Accretion, where neighboring cores compete for any unbound material in their vicinity. The turbulence in the cloud plays a major role here, as cores fly by one another and vie for anything that comes their way. Some amazing animations of this process have been developed and are available on the web at http://www.ukaff.ac.uk/starcluster/. To the casual viewer, the dynamic scenes of chaotic motion and competitive accretion are akin to the skater's game "crack the whip." There are definitely winners and losers! As the animations show, some newborn stars are ejected completely out of the cloud.

As the protostars and their surrounding accretion disks begin to settle down, their spectral emission begins to trend toward higher photon energies and shorter infrared wavelengths. We are beginning to see through the core and pick up the light of the accretion disk itself. The inner part of the dusty disk, being closest to the protostar, is hottest and so emits at the shortest infrared wavelengths. Astronomers can track this evolution in terms of the resulting infrared spectra.

Once the YSO is visible at near-infrared wavelengths, it is deemed to be of spectral Class I. The YSO is still regarded as a protostar whose luminosity is powered solely by the infall and accretion of matter. Some of these YSOs are profuse X-ray emitters as well. This energetic radiation is thought to arise from intense magnetic activity associated with the YSO's accreting disk and outflowing winds. Indeed, one of the best ways to identify YSOs among other stellar sources in a molecular cloud is via their X-ray emission.

Disk Clearing, where the winds and radiation from the YSO remove any residual matter from their vicinity. The winds, in particular, can be extreme, with velocities of several hundred km/s (or *million* km/hr) and mass-loss rates of order 10^{-5} Suns per year—the equivalent of shedding two Earths each year, or a mountain range each second. These intense exhalations cannot go on forever, however, as the star would completely blow itself away in about 100,000 years! In the disk, the congealing of planets can also serve to clear out material. The newly formed planets will both sweep up and disrupt any gas and dust in their orbital paths. The resulting annular gaps in the disk will alter the spectrum of the emergent light. If the gap is close to the star, the hottest material will be absent and so the spectrum will have a deficit of emission at shorter infrared wavelengths. This is observed, providing astronomers with compelling evidence that planets are forming in some YSOs. Once a protostellar system develops these sorts of outflows and odd spectra, it is designated as a Class II YSO.

Thermonuclear Ignition! Meanwhile, the contracting protostar is beginning to look more and more like a star. Strong X-ray emission suggests that it is undergoing convective motions in its interior, with lots of magnetic activity at its surface. The star may even be initiating some thermonuclear fusion of deuterium in its center, as deuterium can "burn" at a relatively low temperature of a few million degrees Kelvin. Astronomers call these more developed protostars *pre-main-sequence (PMS) stars*. The prototype of this class is T-Tauri, which was first identified as being unusual because of its bright and highly variable H-alpha spectral-line emission. T-Tauri stars and their higher-luminosity Herbig Ae/Be pre-main-sequence counterparts have been found in many molecular clouds.

Once the YSO is sufficiently cleared of gas and dust, the PMS star can be observed at optical wavelengths. That means astronomers can plot it on a standard H-R diagram and thereby compare its luminous properties with those of its peers—and to compare these fledgling stellar populations with the bona-fide main-sequence stars that they will eventually become. As shown in figure 6.3, T-Tauri stars are typically brighter than their main-sequence counterparts. That is because the release of gravitational energy during PMS contraction can occur at higher rates than the release of nuclear energy in the subsequent stellar cores.

The Japanese astrophysicist Chushiro Hayashi first predicted in 1961 how PMS stars evolve toward their "adulthood" as hydrogen-fusing main-sequence stars. The so-called Hayashi-Henyey tracks on the H-R diagram show this progression, with lower-mass PMS stars dimming and getting bluer before hitting the main sequence, and with the higher-mass Herbig Ae/Be stars remaining at nearly constant luminosity until they reach the main sequence as blazing blue stars. Typical PMS lifetimes range from less than a million years for the high-mass stars, to tens of millions of years for Sun-like stars, to 100 million years for the lowest-mass stars. Though

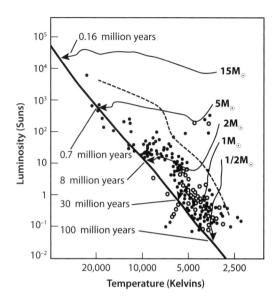

FIGURE 6.3. Hertzsprung-Russell diagram of the young star cluster NGC 2264 showing the evolutionary tracks of pre-main-sequence stars, once they have sufficiently cleared their dusty cloaks. The dashed "birthline" delineates the evolutionary transition from a protostar to a pre-main-sequence star for differing stellar masses. High-mass stars evolve to the Main Sequence hundreds of times faster than Sun-like stars. [Adapted from *Horizons*, 7th edition by M. Seeds, Pacific Grove, CA: Brooks/Cole Publishers (2002), pp. 162, 164]

certainly long by human standards, such lifetimes are small percentages of the H-burning lifetimes that their main-sequence counterparts will enjoy.

The remnant disk that surrounds each PMS star is no longer made of the gas and dust that it once contained. Instead, it is a by-product of rampant collisions between the planets that had formed. This so-called *debris disk* can be seen in several PMS systems, most notably the Beta Pictoris system (see figure 6.4). PMS stars with debris disks are at the last stage of their formative development. With relatively little dust to absorb and re-emit the star's radiation, these systems show no infrared excess in their spectra. Identified by their unusual H-alpha, X-ray, and radio-emitting properties, such PMS systems are designated as Class III YSOs.

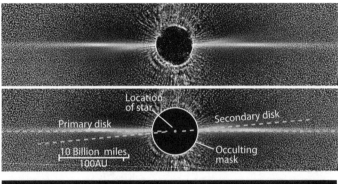

FIGURE 6.4. (*Top*) The debris disk around the pre-main-sequence star Beta Pictoris. This image of the Beta Pictoris system shows the visible light reflected by dust around the young star. Light from the central star has been removed to highlight the circumstellar emission. A bright disk and much fainter tilted disk are evident in this edge-on view. The bright Beta Pic disk is very likely an infant solar system in the process of forming terrestrial planets. (*Bottom*) Artist's rendering of the Beta Pictoris system, complete with debris disk and suspected planets that have been recently confirmed (see figure 4.6). [(*Top*) D. Golomowski, NASA and the European Space Agency. (*Bottom*) Lynette Cook and NASA]

From the Mouths of Babes

Delineation of the evolutionary stages described above is based on a hefty dose of theoretical reasoning as applied to the detailed shapes and motions of molecular clouds and the spectral signatures of discrete sources within them. The evidence for intense outflow activity could not be clearer, however. Optical spectral-line imaging and recent mid-infrared mapping with the Spitzer Space Telescope have shown that many of the swaddled sources in Taurus and other stellar nurseries are squirting jets of ionized gas into their natal clouds (see plate 9). These so-called *Herbig-Haro Objects* are recognized as shocked gaseous projectiles that have been hurled outward from protostars in the thick of accreting gas from their surrounding cloud cores. Some of the accreting gas accumulates in proto-planetary disks that surround the incipient stars, while the rest gets shot out at velocities of several hundred kilometers per second.

Larger-scale outflows have been found in the surrounding molecular gas. In some particularly active star-forming regions, the overall picture becomes an effervescent tableau of infant stars blowing bipolar cavities into their gaseous wombs. We are left wondering what the night sky would look like inside one of these stellar incubators.

Stellar Families

We can also begin to think about what happens when entire families of stars are created wholesale. This is not as unlikely as it sounds. Indeed, most stars are thought to form as part of multiple systems. Some of these systems—including binaries and richer clusters—are obviously bound by their mutual gravity. Others—including the T-Tauri associations found in nearby molecular clouds and the OB associations found farther afield—are more loosely distributed. These newborn stellar populations are of great value to astronomers, because they still retain most of the stars they started out with. That means they provide fully realized snapshots of the stellar demographics that arise in star-forming regions.

Careful measurements of stellar luminosities and spectral types in clustered regions have yielded fairly reliable estimates of each star's mass. Once plotted as a histogram, the frequency distribution of stellar masses shows a characteristic shape (see figure 6.5).

Trapezium Cluster Initial Mass Function

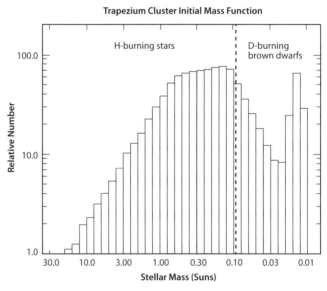

FIGURE 6.5. (*Top*) The Pipe nebula in Ophiuchus the Serpent Bearer, showing distinct complexes of dust clouds and cloud cores in silhouette against the luminous backdrop of stars in the inner galaxy. Characteristic masses of the cloud cores are of order a few solar masses. (*Bottom*) Frequency distribution of stellar masses (the so-called Initial Mass Function) derived from photometric observations of the Trapezium star cluster in the Orion Nebula. Stars of sub-solar mass are most common. This suggests a mass conversion efficiency from cloud core to fully formed star of about 25%. The dotted line indicates the hydrogen-burning limit. Stars of lower mass burn deuterium in their cores. They are known as brown dwarfs. [(*Top*) Yuri Beletsy (European Southern Observatory). (*Bottom*) Adapted from A. Muench, E. Lada, C. Lada, and J. Alves, "The Luminosity and Mass Function of the Trapezium Cluster: From B Stars to the Deuterium-burning Limit," *Astrophysical Journal* 573 (2002): 366–393. Courtesy of C. Lada]

This so-called *Initial Mass Function*, or IMF, tells us the probability of forming a star with a particular mass. Stars with masses between 0.1 and 0.6 that of the Sun appear to be most frequently forged. Some astronomers think that this mass preference is the outcome of the thermal pressures within the star-forming clouds. An equilibrium of sorts is thought to exist between the pressures within the cloud cores and the external pressures exerted by the surrounding inter-core medium. In relatively small molecular clouds like Taurus—or the Pipe Nebula in the direction of the galactic center—the ambient pressures favor the formation of cloud cores with masses of a few times that of the Sun. Allowing for star-forming efficiencies of about 25 percent, such cores should become the natural breeders of stars having masses within the observed sub-solar range.

Nebular Hotspots

So far, we have only considered the star-forming activity that can be seen in relatively modest and quiescent clouds such as the Taurus Molecular Cloud. However, much larger "Giant Molecular Clouds" can be found lumbering around the Galaxy. The masses of these so-called GMCs are estimated to be tens to hundreds of times greater than that of Taurus. Indeed, most of the molecular mass in the Milky Way is thought to reside within the 6,000 or so GMCs that molecular emission-line surveys have revealed. Inside these gaseous behemoths, the likelihood of forming a rare massive star becomes statistically inevitable. And what a difference a massive star makes!

Blazing forth with the power of up to a million Suns, each newborn massive star is a prodigious font of scorching UV photons and scouring winds. An O-type star of 40 solar masses, for example, packs enough of a UV punch to fully ionize and heat up a nebular region several light-years in extent. Indeed, the star's UV radiation alone is sufficient to photo-evaporate the gas, thereby excavating huge cavities into their host clouds. Once cold, dense, and molecular, the newly ionized and expanding gas loudly announces its presence across the electromagnetic spectrum. Astronomers call these stellar-nebular hotspots *HII regions* after the abundant ionized hydrogen (HII) that suffuses them. Many of us (myself included) are drawn to them like moths to a flame. By examining the dynamic inter-relations among the exciting stars and responding gas, we endeavor to understand how HII regions work and what influence they may have on the larger *galactic ecosystems*.

Consider the Orion Nebula, the nearest site of burgeoning star formation—and the best studied by far. Like a crimson rose on a black branch,[1] the Orion nebula has blossomed from the sinuous body of the Orion Molecular Cloud (OMC) complex and is expanding in our general direction (see figure 6.6 and plate 10). The molecular cloud complex takes up much of the constellation of

---

FIGURE 6.6. Views of the Orion constellation and associated star-forming activity. (*Top*) Modern rendering of the Orion constellation with stars and nebulae labeled. (*Middle*) The Orion Molecular Cloud complex, as mapped in the 2.6 mm emission line of carbon monoxide (CO), is recognized as the nearest site of intensive star formation. Much of the clustered star-birth activity is occurring near the belt star Alnitak (where the Horsehead nebula is located) and below the belt stars (where the Orion nebula resides). However, looser distributions of newborn stars have been found throughout the cloud complex, and to the west (right) of the cloud is located a whole association of relatively older B-type stars. Indeed, a progression of eastward trending star formation appears to have transpired over the past 30 million years. (*Bottom*) Far-infrared emission from star-warmed dust in the Orion Molecular Cloud complex. Prominent sources include (*from top to bottom*) a dusty shell surrounding the Lambda Orionis binary star system, the Rosette Nebula(toward the left), the Horsehead Nebula region, and the Orion Nebula region. See also plate 10. [From the website curated by R. Maddalena at http://www.gb.nrao.edu/~rmaddale/Education/OrionTourCenter/opticallinks.html. (*Top*) Chart produced by R. J. Maddalena from the *SkyMap 3.0* program. (*Middle*) Mapped by the 1.2m Columbia University mm-wave radio telescope. Copyright: Public domain. Reference: Maddalena, Morris, Thaddeus, and Moscowitz, *Astrophysical Journal* 303 (1986): 375–391. (*Bottom*) Map generated from FITS files created by the *SkyView* survey analysis system from data produced by the IRAS satellite as analyzed by NASA IPAC/Jet Propulsion Laboratory. Copyright: Public domain. Reference: Wheelock et al., *IRAS Sky Survey Atlas Explanatory Supplement*, 1991]

Visible (stars)

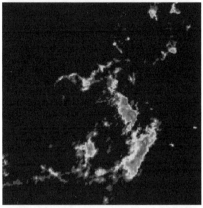

Millimeter-Wave (molecular gas)

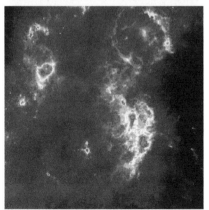

Infrared (dust)

Orion, some $10° \times 15°$ in the sky. Such a large angular extent combined with an estimated distance of 1,500 light-years gives the OMC complex an overall size of 260×390 light-years and a mass equivalent to 500,000 Suns. The fluorescent nebula itself is considerably smaller—of order ten light-years. This modest HII region is still sufficient to host several thousand solar masses worth of newborn stars and ionized gas.

Herein, even amateur telescopes of modest aperture can reveal a rich cluster of stars. The stellar agglomeration becomes especially clear at near-infrared wavelengths, as obscuration by overlying nebular dust is considerably reduced in the infrared. At the core of the cluster reside four massive stars which together are known as the "Trapezium." One of these four stars—Theta 1 Orionis C—is responsible for much of the fluorescent and tumultuous activity in the nebula. Classified as an O6 main-sequence star, Theta 1 Ori C weighs in at about 40 solar masses. Its 35,000-degree surface dishes out prodigious amounts of ionizing UV photons. Indeed, its extreme ultraviolet luminosity is 54 thousand times greater than the Sun's luminosity at all wavelengths. Roughly half of these UV photons ionize and heat the atoms in the nebula until they can recapture their lost electrons and radiate away their newfound energy. This equilibrium between photo-ionization, electron recombination, and subsequent radiation governs the size and luminosity of the nebula. The remaining UV photons irradiate and warm microscopic grains of dust to temperatures of a few hundred degrees Kelvin. Mid-infrared images reveal profligate emission from both the dust grains and the many complex molecules that are present in the Orion star-forming region.

Some of the most bizarre astronomical objects ever found can be seen surrounding Theta 1 Orionis C (see plate 10). Spermlike cometary nebulae show bright irradiated "heads" with "tails" invariably directed away from the hot star. The tails are thought to be flowing off of the denser heads through the photo-evaporative effects of the hot star's UV radiation. Some of the cometary nebulae have *bow shocks* preceding them—testimony to the fierce winds that Theta 1 Orionis C is blowing. Other cometary nebulae contain dark disks of obscuring matter which we now recognize as being proto-planetary in nature. Edge-on and face-on perspectives of these so-called proto-planetary disks or "proplyds" can be seen in silhouette against the background nebulosity. Despite all the tumult that Theta 1 Orionis C has caused, planet-building continues apace. Could our Solar System have been constructed in such a lively nebular environment? We may never know, but if we were born in a rich star cluster, we were likely exposed to the powerful effects of one or two massive stars.

Other more powerful HII regions can be found at greater distances throughout the Milky Way (see figure 6.7 and plate 11). Each of these has its own stories to tell—of massive star birth and its ensuing repercussions. In some instances, the photo-evaporative excavating has left behind great pillars of relatively denser gas. Like the "hoodoos" that rise above the scoured sandstone floor in Bryce Canyon, these gaseous pillars bespeak tales of wholesale erosion and exquisite sculpting. Taking the geological analogy even further, we find ourselves witness to fascinating realizations of the ongoing competition between constructive and destructive agents in nature. As the buildup of mountain ranges causes the weathering and erosion of these same majestic landforms, so too the gestation of stars in molecular clouds leads to the remaining nebulosity being reshaped, rekindled with new star formation, and ultimately destroyed.

Lingering Puzzles

Astronomers continue to struggle with many aspects of the star-birthing process. How long does it take for a dense gaseous core to congeal into a fully formed stellar and planetary system? How do contracting and accreting protostellar systems ultimately shed their contrary magnetic fields and angular momenta? Do massive stars form in ways that are different from their more modest brethren—and if so, how? What happens over time in a large stellar nursery? And how did the first stars form—when there was very

little dust to catalyze the formation of molecular hydrogen, to shield the resulting molecular cores from the disruptive effects of UV irradiation, and to help radiate away the core's excess heat? These are just a few of the vexing questions that keep astronomers searching for telling clues among the fecund clouds of the Milky Way.

FIGURE 6.7. (*Top*) The Eagle Nebula (M16) is a classic example of an HII region, where the intense UV radiation and strong winds from newborn hot massive stars have sculpted and excited their natal surroundings. This image spans about 60 light-years at the distance of M16. (*Bottom left*) Located 6,500 light-years away in the constellation of Serpens the Serpent, M16 has become famous for its "pillars of creation." These protrusions of gas and dust are continuing to be eroded by the hot stars. The tallest pillar in this image measures about 4 light-years. (*Bottom right*) At the tips of the pillars in M16, evaporating gaseous globules (EGGs) have been exposed by the intense UV starlight and winds. The survival of these globules as incubators of future stars and planetary systems remains in question. Elsewhere in the Eagle, star formation continues. [(*Top*) T. A. Rector and B. A. Wolpa (National Optical Astronomy Observatories), Association of Universities for Research in Astronomy, National Science Foundation. (*Bottom*) J. Hester and P. Scowen (Arizona State University), Space Telescope Science Institute, NASA, and the European Space Agency]

CHAPTER 7

LIVES OF THE STARS

I have lived here before, the days of ice,
And of course this is why I'm so concerned,
And I come back to find the stars misplaced
and the smell of a world that has burned.
Well, maybee, maybe it's just a change of climate.

—From "Up from the Skies" in *Axis Bold as Love*

by Jimi Hendrix (1967 CE)

ONE OF THE GREATEST ACHIEVEMENTS of astronomers during the last century was to discover how and then understand why the lives of stars critically depend on the stars' endowed masses. Throughout their "normal" main-sequence phase and more peculiar giant phases, stars illuminate and evolve according to the masses that were allotted them at birth. Only in instances where two closely interacting binary stars exchange mass is this natal determinism compromised. But a star's mass-ordained trajectory of luminosity, color, size, and longevity does not really do justice to its individuality. So before we consider the evolution of stars according to their masses, let's take a moment to marvel at the singular nature of stars.[1]

The Spice of Life

Like snowflakes—and people—each and every star has its own distinct appearance and behavior. This is even true among stars of the same spectral type and luminosity class. Though occupying the same spot on a Hertzsprung-Russell diagram, these stellar look-alikes can manifest individual personalities of fascinating variety.

Main-sequence Sun-like Stars

Consider the Sun and its near twin Alpha Centauri A. Both are classified as G2 main-sequence stars. Their detailed spectra indicate almost identical surface temperatures of 5,780 and 5,770 Kelvins, respectively. The way Alpha Centauri A dances with its cooler companion, Alpha Centauri B, reveals a gravitating mass only 10 percent greater than that of the Sun (see figure 4.2). However, the luminosities of the two stars differ significantly, with Alpha Centauri A being 57 percent brighter. Astrophysicists ascribe this difference to Alpha Centauri A's slightly higher mass and correspondingly more evolved state. As main-sequence stars grow older, the rate of hydrogen fusion in their cores slowly increases. Consequently, the Sun today is 30 percent brighter than it was 4.6 billion years ago, when it first began fusing hydrogen in its core. Although the latest estimates of Alpha Centauri's age are similar to that of the Sun, the star's higher luminosity indicates that it is farther along in its H-burning and so will last only 2 billion more years, compared to the Sun's remaining 5 billion years.

Another key difference between the two stars is the relative amount of heavy elements that spice their mostly hydrogenic interiors. Compared to the Sun, Alpha Centauri A contains 60 percent more iron, sodium, calcium, and other "heavy" elements that have atomic weights greater than that of helium. This compositional contrast provides yet another reason against the notion of the Sun and the Alpha Centauri systems having formed together from the same molecular cloud (see chapter 4). Indeed, both stellar systems may have formed in completely different parts of the Galaxy, but have since meandered into their currently shared locale.

Type A Main-sequence Stars

Shifting our attention up the main sequence to stars of higher luminosity, we cannot help but notice brilliant Vega (Alpha Lyrae) and Sirius A (Alpha Canis Majoris). These blue-white luminaries are respectively classified as A0/A1 main-sequence stars with shared surface temperatures of about 10,000 Kelvins and masses of a little more than twice that of the Sun. Vega is considerably brighter than Sirius A, however, expending 37 solar luminosities compared to a "mere" 25 solar luminosities for Sirius A. In this case, age may be the deciding factor—with Vega clocking in at 400 million years compared to a more youthful 250 million years for Sirius A. Their spin rates are also awry, as Vega twirls around once every half-day while Sirius A takes a full 5.5 days to rotate.

Then there are their disparate surroundings. Deep images and spectra of Vega indicate that it is most likely a single star. As if to compensate for its loner status, it is enveloped in a cloud of dust. The central star irradiates the microscopic dust grains, which then re-radiate the absorbed starlight at far-infrared wavelengths. The dust cloud is thought to be in the shape of a ring as big around as the orbit of Uranus about the Sun. Perhaps we are seeing a remnant debris disk from an episode of planet-forming activity that occurred around Vega shortly after its birth. By contrast, Sirius A has spent much of its life as a binary companion to an even more massive and luminous star (see figure 5.4). This binary relation might explain why Sirius A rotates much slower than Vega, as the binary orbit between Sirius A and Sirius B would have used up much of the stellar system's primordial angular momentum. Sirius B may once have been a hot class B3 star that could have contained as much as six or seven solar masses. It has since evolved completely through its main-sequence and giant phases—exhaling more than 80 percent of itself to become a remnant white dwarf. We will consider these bizarre leftovers of stellar evolution in the next chapter.

In our brief comparison of A0 main-sequence stars, we would be remiss to neglect Cor Caroli (Alpha Canum Venaticorum)—a true oddball among the genre. Situated a few degrees south of the Big Dipper's handle, Cor Caroli is a binary system consisting of the F0 class star Alpha-1 and the A0p class star Alpha-2. Here, the "p" designation refers to the peculiar properties of the A0 star, as Alpha-2 is decidedly weird. Detailed spectroscopy of Alpha-2 has shown it to be a "magnetic star," with surface magnetic fields that are 1,500 times stronger than that on the Earth or the average field on the Sun (away from sunspots). This intense magnetic field appears to have dredged up some heavy elements from the star's interior and submerged others—creating strange enhancements of silicon, mercury, and even rare europium at its visible surface. The upshot of all this activity (plus a slightly greater mass of 2.8 times that of the Sun) is a much higher luminosity that is equivalent to the radiant output of eighty-three Suns. In its deviant behavior, Cor Caroli reminds us of the limitations to our sweeping generalizations based on basic observational classifications.

Extreme Main-sequence Stars

The verity of our stereotyping is stretched even further when we consider the dimmest red orbs at the bottom end of the main sequence and the most luminous stars at the uppermost blue end. Down in the low-luminosity cellar, there are many M-type stars to gaze upon. Their spectra are rich with absorption lines of calcium, sodium, iron, and other heavy metals, as well as absorption bands from titanium oxide, vanadium oxide, and other molecules. Indeed, this spectral richness gives rise to all sorts of variety which is difficult to pin down. Then there is the flaring. Some M-type stars show occasional emission in the lines of crimson H-alpha, teal H-beta, and other colorful atomic transitions. For example, the nearest star to us, Proxima Centauri, has been observed to suddenly brighten in these lines. Classified as a M5.5e star (the "e" denoting its flaring status), Proxima Centauri is likely manifesting a complex and unstable magnetic field which occasionally "collapses." The resulting release of magnetic energy can literally double the star's luminosity. Among the thirty-six M-type main-sequence stars within fifteen light-years of the Sun, seventeen are known to go through these sorts of flaring histrionics.

At the brightest end, where the O-type stars blaze forth, the number of spectral lines is minimal—with only neutral and singly ionized helium evident in absorption. From these lines, astronomers can get a good idea of the surface temperatures, but that's about it. The relative strengths and individual widths of the Doppler-broadened absorption lines yield characteristic temperatures of 30,000–50,000 Kelvins. Any other spectral details resulting from the star's particular surface gravity, rotation rate, and magnetic field configuration are pretty much overwhelmed by the Doppler broadening. The other problem is finding enough bona-fide main-sequence O-type stars, as these massive stars are rarely ever made and evolve away from the main sequence so quickly. Within only a few million years, one is dealing with windy supergiant O- and B-type stars and even more eruptive Wolf-Rayet stars—where bubbles of hot gas are violently expelled in ways that are unique to each star.

Giant and Supergiant Stars

Individual personalities also abound among the cooler giant and supergiant stars. This is especially true for the giants and supergiants whose temperatures and luminosities place them on the Hertzsprung-Russell diagram somewhere along the so-called instability strip (see figure 7.1). Here, variability is the rule, as the stars literally heave outward and inward in repetitive cycles.

The cause of these periodic pulsations lies within a layer deep beneath the surface of each star. The flux of ultraviolet photons and temperature of the gas in this layer are "just right" for the helium atoms to alternate between states of single and double ionization. When singly ionized, the helium is primed to absorb the ultraviolet photons from below and become doubly ionized. The liberated electrons add kinetic energy to the gas, thus producing higher temperatures and enhanced pressures within the layer. As the energized layer pushes outward, the outer layers of the star react by expanding to greater dimensions. The expansion, in turn, dilutes the energy density of the gases, reducing the temperatures, pressures, and ionization states in the layer back toward their former state. The deflated star contracts under the inexorable force of gravity—priming itself for a renewed cycle of energy absorption, pressurization, and expansion.

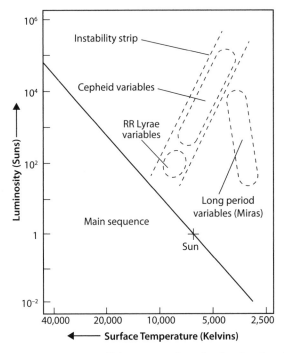

FIGURE 7.1. A Hertzsprung-Russell diagram with stellar luminosity plotted against surface temperature. The so-called instability strip runs from the Main Sequence upward toward higher luminosities and cooler temperatures. It includes RR Lyrae variables and different types of Cepheid variable stars. Long-period (Mira) variables occupy their own part of the H-R diagram. [Adapted from *Discovering the Essential Universe*, 3rd edition by N. F. Comins, New York: W. H. Freeman (2006): 255]

Cepheid Variable Stars

The most famous of the variable stars are the Cepheids, named after the prototype Delta Cephei. This F5Ib class yellow-white supergiant dishes out an average of 2,000 solar luminosities from a surface that has swollen to forty solar diameters—half the size of Mercury's orbit around the Sun. It can be seen naked-eye, pulsating by a factor of two in brightness (0.8 magnitudes) over an amazingly regular period of 5 days, 8 hours, 47 minutes, and 32 seconds (see figure 7.2). As it varies in brightness, its radius changes by only a few percent. Its surface temperature, however, changes from 6,800 Kelvins at its warmest to 5,500 Kelvins when it is coolest and dimmest. This is enough to alter its spectral type from F to G over the same time interval.

The other members of the Cepheid class undergo similar upheavals, brightening and fading over time in signature "sawtooth" patterns. Moreover, their periods of pulsation are directly correlated with their average luminosities, such that the more luminous Cepheids display the longer periods (see figure 2.13). This so-called period-luminosity or P-L relation has become the gold standard for determining distances throughout the local universe. Wherever a Cepheid variable can be found and monitored, its pulsational period can be determined and its average luminosity then inferred from the P-L relation. The star's intrinsic luminosity can then be compared with its apparent brightness to obtain the distance to the hosting system. Using Cepheid variable stars as "standard candles," astronomers have fathomed distances throughout the Milky Way, the Local Group of galaxies, and more remote galaxy groups out to the Virgo Cluster of galaxies some 60 million light-years away.

As you might expect, the "instability strip" also plays host to a dizzying number of stars that defy such straightforward typecasting. For example, Mirfak (Alpha Persei) shares the same F5Ib classification as Delta Cephei. Yet it is 2.5 times more powerful, with a mean luminosity of 5,000 Suns, and is considerably larger with a girth that is 62 times greater than that of the Sun. It is also far more steadfast than Delta Cephei, with any pulsations below the limits of perceptibility. Apparently, the interior conditions within Mirfak cannot support the sustained oscillations that characterize the classic Cepheid genre. The star's higher luminosity and warm surface temperature may play a role here, as these properties place it on the Hertzsprung-Russell diagram near the upper margin of the instability strip.

Mirfak's current state indicates to astronomers that it was most likely an 8 solar-mass B-type main-sequence star about 30–50 million years ago. Today, it is the crowning jewel of the Alpha Persei cluster about 590 light-years away. The cluster once hosted even more massive stars, but they have long since expired in violent

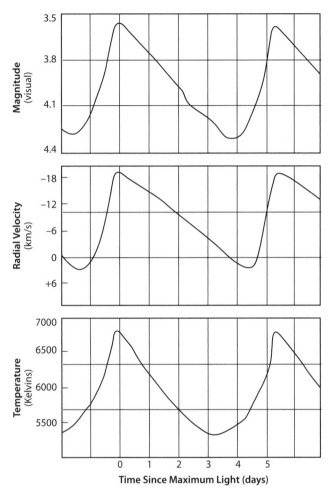

FIGURE 7.2. Delta Cephei shows periodic variability in its light output, pulsation velocity, and surface temperature. Other Cepheid variable stars show similar behavior. Moreover, their pulsational periods correlate with their average luminosities—a key relation that can be used to determine distances throughout the Milky Way and beyond to other galaxies. [Adapted from *Burnham's Celestial Handbook* by R. Burnham, Jr., Mineola, NY: Dover Publishers (1978), p. 585]

supernova explosions. Perhaps Gould's Belt of current star-forming activity was shaped and triggered by these historic explosions (see chapter 4).

Other Cepheid variables seem to be changing before our very eyes. Polaris, the "North Star" (Alpha Ursae Minoris), is most notable, as it is often considered the most "constant" of stars. In truth, Polaris is a F71b class yellow supergiant that is varying in brightness over a small amplitude of 0.03 magnitudes (about 3 percent). Its pulsational period of 3.97 days is anomalously short for its luminosity of 2,500 Suns. These modest and disparate pulsations appear to be diminishing even further. Astronomers suspect that Polaris is currently vibrating at its first "overtone" frequency and that it will soon transition back to its "fundamental" beat, with a longer 5.7-day period and much greater amplitude. Best to stay tuned!

Red Supergiants

We end our brief homage to stellar individuality by comparing the properties of Betelgeuse and Antares—the two most dazzling red supergiants in the naked-eye sky. Both stars share the ruddy hues characteristic of their M1.5 spectral types and corresponding surface temperatures of 3,600 Kelvins. They also share the special status of being among the largest stars known. Indeed, they are so large and sufficiently close that their surfaces can be directly imaged. Betelgeuse (Alpha Orionis) ranks as the more extreme leviathan, with a size of 1,200 solar diameters. Were it placed at the Sun, the star would overfill Jupiter's orbit by 20 percent. Accounting for its prodigious infrared output yields a total (bolometric) luminosity that exceeds 60,000 Suns. Slowly varying in brightness over multiple periods of ½ to 6 years, Betelgeuse is exhaling a strong wind that has created a shell of gas and dust around it.

Antares (Alpha Scorpii) is a tad smaller, with a diameter that spans "only" 760 Suns. Because it is such a strong infrared emitter, the star's bolometric luminosity rivals that of Betelgeuse. Antares also varies over a semi-regular period of several years and is blowing a fierce wind into space. Some of these exhalations have collected in the form of a circumstellar nebula that shines by the scattering of light from the central star. Coursing through the morass is a blue-hot companion star of B2.5 spectral type. Its UV-bright radiation has created a small ionized region within the dusty nebula. Meanwhile, both Antares and Betelgeuse are racing to the end of their respective lives. Given their individual endowments of about fifteen solar masses, these stellar behemoths have been profligately blazing for several million years and have pretty much exhausted their thermonuclear reserves. One or both of these stars will likely explode within the next few hundred thousand years. That virtually guarantees an incredible show for some hominid species—perhaps even our own.

Inner Workings

Built up by gravity's inexorable attraction, and powered by the intense conditions that this self-gravity has produced in their cores, stars can be regarded as finely tuned furnaces. Astrophysicists have made tremendous progress in understanding how these gaseous orbs sit upon themselves and so induce the extraordinary central pressures and temperatures that are required for thermonuclear reactions to proceed. The resulting flood of gamma rays is what energizes the star and keeps it from gravitationally imploding, even though it is radiating away copious amounts of energy into space. This balance between the star's inward pulling gravitational force and outward pushing thermal pressure is known as "hydrostatic equilibrium" (see figure 7.3). Fortunately for us all, the equilibrium is a stable one. Any forced compression or expansion of the star will alter the gravitational and thermal energies, so that the star reverts back to its initial state.

Stars of different mass perform their balancing acts in different ways. In the following descriptions, we must rely on the self-consistent models that astrophysicists have devised. These have turned out to be highly successful, explaining—and often predicting—the behavior of both main-sequence and giant stars over an impressive range of masses.

Main-sequence Stars of Low Mass (0.08–0.8 Solar Masses)

Within this range of masses, the stars are thought to be completely convective (see figure 7.4). That means there is always fresh hydro-

Stellar Surface

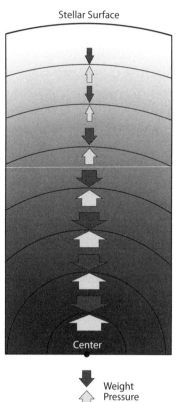

FIGURE 7.3. Cartoon of the balance between inward-pulling gravitation and outward-pushing thermal pressure. Each layer in the star is balanced in this way, producing a state of "hydrostatic equilibrium." Because of the tremendous weight above them, the innermost layers are the most pressurized. Near the center of the Sun, for example, the corresponding pressures are a billion times greater than that at the bottom of Earth's oceans. The resulting densities of 156 grams/ cm³ and temperatures of 15 million Kelvins in the Sun's core are sufficiently high to permit the thermonuclear fusion of hydrogen into helium. [From Seeds, *Horizons: Exploring the Universe (with The Sky CD- ROM and InfoTrac)*, 7E. © 2002 Wadsworth, a part of Cengage Learning, Inc. Reproduced by permission. www.cengage.com/ permissions]

gen available to feed the thermonuclear reactor in the star's core. The reaction likely involves the "proton-proton chain" that binds 4 hydrogen nuclei (4 protons) into one helium nucleus (2 protons and 2 neutrons) plus 2 neutrinos and 2 gamma rays.[2] The mass of the 4 hydrogen nuclei is 0.007 times greater than that of the resulting helium nucleus (I remember this by thinking of James Bond—a stellar powerhouse of another sort). This mass excess is converted into radiant energy according to Albert Einstein's famous equation

$$E = mc^2,$$

where in this case the energy E is in the form of gamma rays, the excess mass m is 0.007 times the mass available in the form of hydrogen nuclei, and the speed of light c is a very large number

(300,000 km/s). Once this latter quantity is squared and multiplied by the excess mass, a tremendous amount of energy can ensue.

The gamma rays from the thermonuclear core are immediately absorbed by the overlying gases which then heat up, expand, and buoyantly float up to the surface. Radiating away their thermal energy into space, the surface gases readily cool off and become more susceptible to falling back into the deep interior. This sets up a full-scale convective flow that involves the entire star. Because low-mass stars are so completely convective, they can continue fusing hydrogen in their cores throughout their lives as stars. And these lives are long indeed—much longer than the universe has been in existence! Every low-mass star that the Milky Way has ever spawned is still around today.

Main-sequence Stars of Intermediate Mass (0.8–8.0 Solar Masses)

Stars like the Sun are also powered by the proton-proton chain of thermonuclear reactions. Where they differ from their lower-mass brethren is in how the energy gets transported to the surface (see figure 7.4). Between 0.8 and 1.2 solar masses, the thermonuclear core of the star is surrounded by a radiative zone. Here, the out-flowing gamma rays are absorbed by the overlying ions and electrons, which then heat up to a temperature somewhat lower than the core's temperature of 15 million Kelvins. The heated layer re-radiates photons of lower X-ray energies consistent with its lower temperature. The X-ray photons, in turn, are absorbed by the next higher layer, which then warms and re-radiates at even lower ultra-violet energies. Like middle-men turning a $100 bill into 10,000 pennies, the particles in the radiation zone convert the original flux of gamma rays into a flood of UV and optical photons. The photons careen off the charged particles in the plasma in a sort of drunkard's walk that has a slight but consistently outward direction—compelled by the decreasing radial gradient in temperature. The whole energy transport process can take several million years! Once the radiant energy is about two-thirds of the way to the surface, roiling convection takes over—transporting the heated gases to the surface, where they cool and sink back in a continuing cycle.

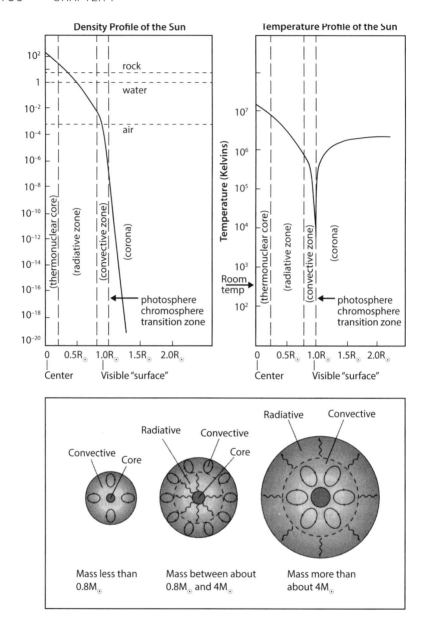

FIGURE 7.4. (*Top*) Radial profiles of density and temperature in the Sun, with the radiative and convective zones marked. (*Bottom*) Schematic interiors of low, intermediate, and high-mass stars, showing the differing ways in which energy is transported outward. [(*Top*) W. H. Waller and L. Slingluff based on multiple sources. (*Bottom*) Adapted from *Astronomy—CliffsQuickReview* by C. J. Peterson, Foster City, CA: IDG Books Worldwide, Inc. (2000), p. 164]

Above 1.2 solar masses, energy generation in the core is provided by the so-called CNO process. Trace amounts of carbon, nitrogen, and oxygen nuclei catalyze the fusing of hydrogen into helium (along with an enhancement of nitrogen at the expense of carbon and oxygen). This reaction requires higher central temperatures and densities. The payoff is a much greater reaction rate which drives convection in the lower layers. Above the inner convective zone, radiative processes transport the energy the rest of the way to the star's surface.

Main-sequence Stars of High Mass (8.0 to about 120 solar masses)

Life gets ever more dramatic for stars with masses above eight solar masses. The crushing weight of these ponderous stars compresses their cores to incredible pressures and temperatures. The resulting thermonuclear reactions go into overdrive, spewing torrents of gamma rays that challenge the stars' ability to cope.

We have seen that lower-mass stars achieve hydrostatic equilibrium by balancing their thermal pressures against gravity in each and every layer. For the higher-mass stars, a new agent of contention enters the fray. The emerging photons themselves transport enough mechanical energy to push on the overlying layers and destabilize them. A star that exceeds fifty solar masses or so really begins to feel the effects of this photonic pressure. It becomes susceptible to blowing itself away, even before it is fully formed. The British astrophysicist Sir Arthur Eddington was the first to recognize the importance of "radiation pressure" in high-mass stars. In the 1920s, he determined that there is a maximum luminosity, above which no star can keep itself together. This amounts to about 20 million solar luminosities and a corresponding mass of about 120 Suns. So far, we have yet to find a star whose luminosity and mass significantly exceed this so-called Eddington limit. Instead, we find stars of spectral class O3 and estimated masses of 80–150 Suns that are rapidly evolving into hot blowhards.

Evolutionary Tales

As a star evolves beyond its hydrogen-fusing main-sequence phase, its mass once again determines what happens next. Of course, it's all a losing battle against gravity, as every star eventually runs out of ways to keep the central fires burning and hold off the inevitable crunch. Stars of intermediate mass are able to keep things going in several clever ways for billions of years. Higher-mass stars rely on fusing ever heavier elements in a last desperate attempt to hold off the inevitable implosion. In less than 10 million years they succumb—dying in spectacular supernova explosions.

Most stars spend 80–90 percent of their lives steadily fusing hydrogen into helium. This is done via the proton-proton chain or—for the more massive stars—the CNO cycle. Even in their relatively stable phase of life, irreversible change is afoot. As hydrogen nuclei transmute into heavier helium nuclei, fewer numbers of particles are available to zip around and collide with one another. One can get a clue as to the consequences by considering the *Ideal Gas Law* which relates the pressure and temperature of a gas whose particles are undergoing perfectly elastic collisions with one another. This law states that the pressure is directly proportional to the temperature according to

$$P = nkT,$$

where P is the pressure, n is the number density of particles (number per unit volume), k is the Boltzmann constant, and T is the temperature.[3] The ongoing fusion of four hydrogen nuclei into one helium nucleus results in lowering the total number density of particles. In order to maintain sufficient thermal pressure and so prevent gravitational collapse, the core must readjust itself to generate higher temperatures. This entails boosting the thermonuclear reaction rate and resulting luminosity.

The star's changing luminosity and surface temperature can be plotted on a Hertzsprung-Russell diagram. For a star like the Sun, the so-called evolutionary track involves a doubling of luminosity over the star's 10 billion-year main-sequence lifetime (see Figure 7.5). As the Sun was significantly dimmer in its youth, so too will

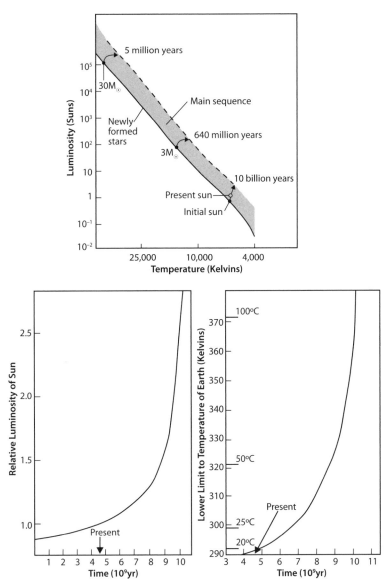

FIGURE 7.5. (*Top*) Hertzsprung-Russell diagram showing estimated evolutionary tracks of stars while on the Main Sequence. Stars like the Sun take a full 10 billion years to evolve through its main-sequence stage of core hydrogen fusion into helium. Massive stars take less than a thousandth of that time to burn through the hydrogen in their cores. (*Bottom*) Resulting changes in the Sun's luminosity and the Earth's surface temperature over time. [(*Top*) From Seeds, *Horizons: Exploring the Universe (with The Sky CD- ROM and InfoTrac)*, 7E. © 2002 Wadsworth, a part of Cengage Learning, Inc. Reproduced by permission. www.cengage.com/permissions. (*Bottom*) Adapted from *Atoms, Stars, and Nebulae*, 2nd edition by L. A. Aller, Cambridge, MA: Harvard University Press (1971), p. 213]

it be considerably brighter in a few billion years. The effect on Earth will be severe. The Earth's surface will continue to heat up until the oceans boil away and life as we know it becomes impossible. Our current worries over global warming are mere trifles compared to these prospects!

Giant Phases

Once hydrogen in the thermonuclear core has been completely converted into helium, the star manages to keep itself from imploding by letting its core contract while fusing hydrogen in a shell just beyond the core. Numerical simulations have been able to follow this dynamical process in great detail. They have shown that the reconfiguration of energy generation results in an overall inflation of the star. A Sun-like star will fluff up until it is the equivalent size of Mercury's orbit. The star is now known as a *red giant*, befitting its larger, redder, and brighter state (see figure 7.6).

Eventually, the core contracts enough for helium to begin fusing into carbon and oxygen. This is carried out via a reaction that requires higher temperatures (about 100 million Kelvins) and yields much less energy per reaction. However, the rate of the so-called *triple-alpha process* is so high that it more than makes up for the lower yield per nucleon. The star, now in its *horizontal branch* phase, is able to sustain these profligate ways for only about 10 percent of its former main-sequence lifetime.

Sometime during this period, the star will undergo oscillations similar to those of Cepheid variables, but at much lower luminosities. We recognize such variable stars as RR Lyraes—whose nearly constant average luminosities provide helpful "standard candles" for gauging distances to evolved star clusters in the Milky Way. All too soon, the star has completely burned through its central supply of helium. The core goes dormant again and re-contracts, while helium burning ensues in a shell just outside the core and hydrogen burning continues in a shell just beyond the helium fusing shell.

All this shell burning boosts the star's luminosity and so creates a distinctly recognizable track on the Hertzsprung-Russell diagram known as the *asymptotic giant branch*. A Sun-like star in this state

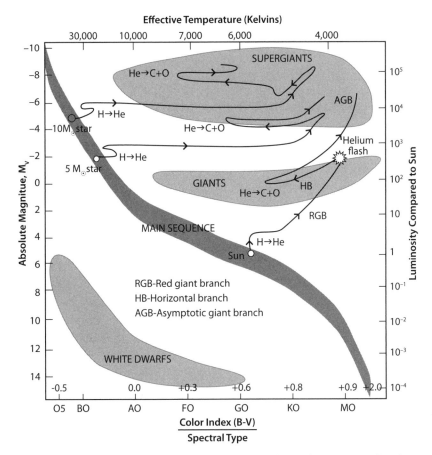

FIGURE 7.6. Hertzsprung-Russell diagram showing the evolutionary tracks of intermediate-mass stars during their red-giant, horizontal-branch, and asymptotic-branch phases. [Adapted from the Australia Telescope Outreach and Education website at http://outreach.atnf.csiro.au/education/senior/astrophysics/images/stellarevolution/hrpostmainsuntrack.jpg, hosted by the Commonwealth Scientific and Industrial Research Organization (CSIRO) of Australia]

would be as big as the orbit of Mars! When the Sun itself reaches this stage in its life, the Earth will likely no longer exist—having vaporized within our bloated star's distended atmosphere, or having been flung from the Solar System long before. Instabilities within the star lead to cycles of alternating helium and hydrogen fusion in their respective layers. The resulting thermal pulses and their irradiation of the star's dusty outer layers can drive off tremendous winds, creating circumstellar nebulae and exposing the hot, dense interiors. We will consider these stellar and nebular exotica in the next chapter.

Supergiant Phases

For more massive stars, central pressures and temperatures are sufficiently high to continue the process of fusing ever heavier elements in their cores and sustaining multiple shells of fusing elements (see figure 7.7). With each conversion of one element to a heavier element, the number density of particles decreases. That means the core must burn through its new supply of fuel with ever increasing haste in order to hold off imminent death by gravity. Consider a 25 solar mass star. As shown in table 7.1, such a barn-burner takes about 7 million years to consume its core hydrogen, 500 thousand years to convert the resulting helium to carbon and oxygen, a half-year to fuse the oxygen into silicon, and only a day to transmute the silicon into iron.

During these various power plays, the massive star's surface temperature wildly swings from being red-cool to blue-hot and back, while its total luminosity creeps upward. At times, the star's luminosity and temperature place it on the "instability strip," where it undergoes regular pulsations as a Cepheid variable. All too soon, it hits the end of the line—with an iron core that obstinately refuses to fuse into anything heavier. That is because the iron nucleus is the most tightly bound of all atomic nuclei (see figure 7.7). Any further fusion reactions require more energy than they release. Indeed, all elements heavier than iron are more likely

FIGURE 7.7. (*Top*) Onion-skin structure deep inside a 25 solar-mass star just before its iron core implodes. Each elemental shell in the star's interior is undergoing fusion reactions that briefly keep the star pressurized against wholesale gravitational collapse. (*Bottom*) Curve of nuclear binding energy as a function of the atomic weight. These energies are negative in the sense that the fusion of lighter nuclei into heavier nuclei of greater negative binding energy involves the release of positive energy that heats and pressurizes the star. Nuclei of iron (Fe) have the greatest negative binding energies, and so must *consume* positive en-

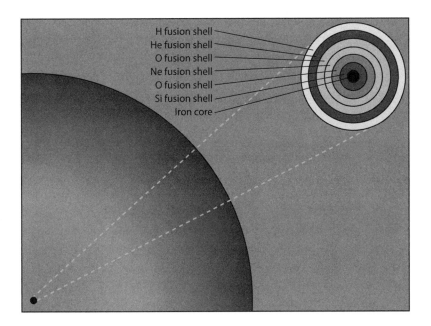

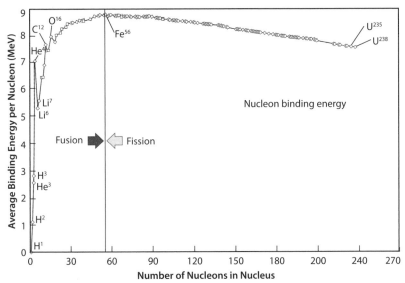

ergy to fuse into heavier nuclei. That's why massive stars stop fusing elements once they have made iron in their cores. [(*Top*) From Seeds, *Horizons: Exploring the Universe (with The Sky CD- ROM and InfoTrac)*, 7E. © 2002 Wadsworth, a part of Cengage Learning, Inc. Reproduced by permission. www.cengage.com/permissions. (*Bottom*) Adapted from Wikipedia Commons (see http://en.wiki pedia.org/wiki/File:Binding_energy_curve_-_common_isotopes.svg)]

TABLE 7.1. Fusion Timescales for a 25 Solar-mass Star

Fuel	Core Temperature (Kelvins)	Fusion Timescale	Percentage of Total Lifetime
Hydrogen (H)	40 million	7 million years	93.3
Helium (He)	200 million	500 thousand years	6.7
Carbon (C)	600 million	600 years	0.008
Neon (Ne)	1.2 billion	1 year	0.000013
Oxygen (O)	1.5 billion	6 months	0.000007
Silicon (Si)	2.7 billion	1 day	0.00000004
Iron (Fe)	5.4 billion	No fusion → Core collapses in 0.2 second	

to decay by fission than to fuse into more ponderous species. The newly dormant core, unable to maintain its thermal pressure, implodes in less than a second.

The resulting release of gravitational energy is stupendous. Neutrinos stream out of "ground zero" and interact with the rest of the star. The ensuing shock waves ripple through the star—detonating a supernova explosion of titanic proportions. For the next weeks to months, the supernova can outshine the entire galaxy in which it resides. Much of this afterglow is the result of elements heavier than iron that were forged in the supernova's intense fires. As nickel and other heavy elements decay into lighter isotopes, they contribute to the supernova's overall radiance in a characteristic way (see figure 7.8). Through observations of this decay process, we are witness to true cosmic alchemy. Indeed, we can attribute the entire bottom half of the periodic table of elements to these sorts of fiery origins.

Supernova explosions are thought to occur somewhere in the Milky Way every thirty years or so. Most will be obscured by the dust in the disk, but even considering this effect, we seem to be overdue for a big one. The last visible supernova in the Milky Way was recorded by Johannes Kepler in 1604. We have to look beyond

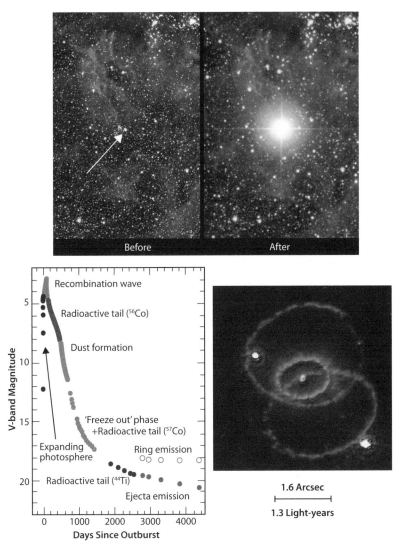

FIGURE 7.8. (*Top*) Before and after views of the region in the Large Magellanic Cloud, where Supernova 1987A exploded. (*Bottom left*) Light curve of SN1987A shows slowly declining emission powered by the radioactive decay of nickel-56 into cobalt-56 and other heavy elements. (*Bottom right*) Circumstellar rings around SN1987A illuminated months after the initial explosion. [(*Top*) David Malin (Australian Astronomical Observatory). (*Bottom left*) European Southern Observatory. (*Bottom right*) Hubble Space Telescope (Space Telescope Science Institute), NASA, and the European Space Agency]

our Galaxy to note anything more recent. Supernova 1987A in the Large Magellanic Cloud was especially important, as we could detect the first neutrinos that streamed forth. We were also able to track the subsequent evolution of the supernova's afterglow in great detail. The progenitor star was previously classified as a hot blue supergiant of spectral type B3 I and estimated mass of 20 Suns. It had earlier vented much of itself in a strong wind whose effluvia collected around the star in the form of a nebular "hourglass." When the star exploded, its burst of light had to travel several light-months before striking the gases near the surface of the hourglass. Astronomers recorded this event as a series of circumstellar rings months after the main event (see figure 7.8). Over the past twenty years, astronomers have monitored the rings, tracking their brightening as sonic shock waves from the supernova explosion finally arrived.

In the next chapter, we will consider what gets left behind after stars of differing mass reach the end of their nominal lives. These exotic stellar and nebular remnants play vital roles in the ongoing saga of the Milky Way, as will be explored in the subsequent chapters.

CHAPTER 8

• • • ● • • •

STELLAR AFTERLIVES

Fire all your guns at once and—explode into space.
—From "Born to Be Wild" by Mars Bonfire
for Steppenwolf (1968 CE)

I believe a leaf of grass is no less than
the journey-work of the stars.
—From *Leaves of Grass*
by Walt Whitman (1900 CE)

NOTHING LASTS FOREVER, not even stars. In their deaths, stars leave behind fleeting nebulae of diaphanous beauty along with much longer-lasting compact remnants whose bizarre properties defy our Earth-bound imaginings. These nebular and stellar exotica play peculiar but vital roles within their hosting galaxies. Once again, stellar mass determines what ensues. We can see in figure 8.1 a road map of the different legacies that result from differing stellar masses. From white dwarfs, to neutron stars and black holes, myriad weird residues of stellar lives can be found populating the Milky Way's disk and halo. Meanwhile, the enriched effluvia of once magnificent stars have infiltrated our Galaxy's interstellar medium—spicing future sites of star formation and making possible the genesis of rocky planets along with the carbon-based life that has emerged on the moist surface of one such world. Thanks to the generosity of the many stars that have gone before us, we can truly claim to be "starstuff"—as connected to the cosmos as the galaxy that spawned us.

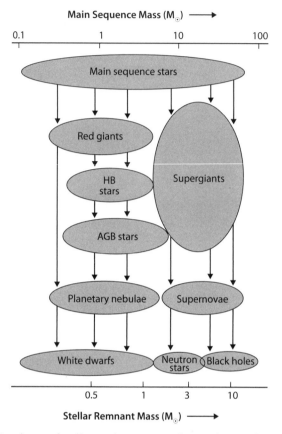

FIGURE 8.1. Roadmap of stellar evolution, parsed according to the star's initial mass (*top*) and stellar remnant's final mass (*bottom*). [Adapted (with changes) from *Discovering the Universe*, 4th edition by W. J. Kaufmann and N. F. Comins, New York: W. H. Freeman (1996)]

Fates of Sun-like Stars

We don't really know what happens to stars of very low mass, as the universe hasn't been in existence long enough for any low-mass star to die. Some theorists think that such a star can keep itself sufficiently churned up so that fresh hydrogen fuel continually feeds its thermonuclear core. Once all the hydrogen has fused into helium, the star could eventually collapse into some sort of Jupiter-like remnant whose thermal pressure can oppose its self-gravity well into perpetuity. Many other astrophysicists think another fate awaits these stellar lightweights. Without a central furnace to keep the star toasty and pressurized, it will continue to collapse until it forms a helium-rich *white dwarf*. Such a compact remnant would be similar to the more massive (and more familiar) carbon-oxygen white dwarfs—the condensed residues of intermediate-mass stars like the Sun.

Planetary Nebulae

For stars with masses between 0.8 and 8.0 Suns, the end state gets decidedly peculiar. Having previously segregated their interiors into thermonuclear, radiative, and convective zones, these stars have left their cores to fend for themselves. The upshot has been a sequence of core contractions accompanied by the fusion of hydrogen in the surrounding shell, ignition of helium fusion in the core, and then shell-burning of both helium and hydrogen. Unable to spark fusion of the carbon-oxygen mix in its core, the star ends its thermonuclear career as a double-shell burner. Thermal pulses from the alternating sources of fusion drive the star to ever higher luminosities. The resulting profusion of photons impinges upon any dust grains that may have crystallized in the star's cooler outer layers. Pushed outward by the photonic pressure, the motivated dust grains drive an overall wind of colossal proportions. The star's liberated layers fly away at speeds of hundreds of kilometers per second, forming a circumstellar nebula made of shells, plumes, curlicues, and all sorts of other fantastical shapes. Meanwhile, the hot core of the star is exposed to view for the first time. Torrents of ultraviolet photons stream outward, many of which impact the atoms in the nebula. These ionize the atoms and excite the resulting ions of hydrogen, helium, nitrogen, sulfur, and oxygen to fluoresce in their characteristic spectral emission lines. With each excited element, a unique palette of hues and intensities is added to the spectral mix.

What we see through our eyepieces and in our deepest photographic exposures are delightful shapes of fuzzy luminescence (see figure 8.2 and plate 12). William Herschel, the discoverer of Ura-

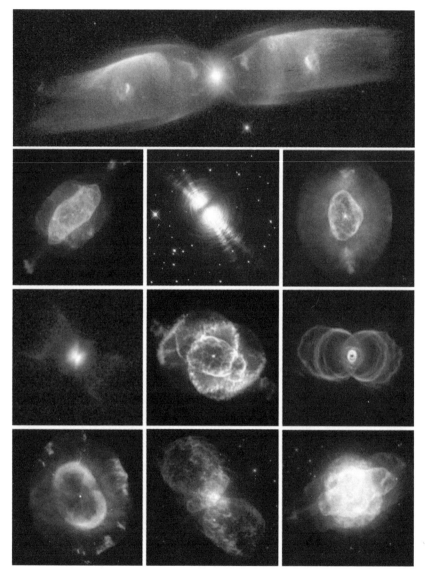

FIGURE 8.2. Gallery of planetary nebulae, as imaged by the Hubble Space Telescope. (*Top row*) Butterfly Nebula (M2-9). (*Second row*) Saturn Nebula (NGC 7009), Egg Nebula (CRL 2688), and Ghost of Jupiter (NGC 3242). (*Third row*) Red Rectangle, Cat's Eye Nebula (NGC 6543), and Hourglass Nebula (MyCn 18). (*Bottom row*) NGC 7662, Double Bubble Nebula (Hubble 5), and NGC 3918. [B. Balick (University of Washington), H. Bond (Space Telescope Science Institute), R. Sahai (Jet Propulsion Laboratory – Caltech), their collaborators, NASA and the European Space Agency]

nus, dubbed these celestial objects "planetary nebulae" based on their small sizes and greenish tints as viewed through his giant reflecting telescopes. We now know that planetary nebulae owe their shapes to the stellar winds that have blown, and that these winds can be shaped by any preexisting factors such as the presence of an impeding debris disk, a gravitating binary companion—or a substantial planetary system. So the connection between planetary nebulae and planets is today recognized as being far less direct but certainly plausible.

The planetary nebula stage is incredibly brief—lasting only a few tens of thousands of years. Our ice ages come to mind. Because of their ephemeral nature, one might think that planetary nebulae would be exceedingly rare. However, there are lots and lots of progenitor intermediate-mass stars in the Milky Way—of order 10 billion (10^{10}). Even if you cull this number according to the ratio of lifetimes between the corresponding planetary nebula and main-sequence phases (about 10^4 yrs/10^{10} yrs), that still leaves roughly 10^4 planetary nebulae currently adorning the Galaxy. To date, astronomers have found about 1,500 planetary nebulae in the Milky Way—well below the estimated number. Most likely, the others are too far and faint, or too obscured by dust in the disk to be detected.

Though short-lived, the planetary nebulae provide vital venues for enriching the Galaxy. Whatever the progenitor stars cooked up in their thermonuclear cores (helium, carbon, nitrogen, and oxygen) are exhaled along with whatever elements these stars had inherited from prior stellar generations (including sulfur, silicon, and iron). As these diverse elements recede ever farther from their stellar hosts, they cool enough to chemically bond with one another. The result is a wizard's brew of silicaceous and carbonaceous dust grains—each microscopic mote surrounded by a "mantle" of various ices (H_2O, CO_2, NH_3, etc.). Within the dusty haze are all sorts of complex organic compounds—including polycyclic aromatic hydrocarbons (PAHs) based on the benzene ring (C_6H_6), various fragrant chains of hydrocarbons worthy of Chanel No. 5, and perhaps even cursory amino acids such as glycene. Several of these biochemical precursors have been detected in surprising abundance through their characteristic emission at radio and infrared wavelengths. In our observations of dying giants and planetary nebulae, we are witness to nothing less than the chemistry of life in its earliest incarnations (see plate 12).

White Dwarfs

Meanwhile, the exposed stellar cores are once again left to their own devices. Bereft of thermonuclear support, these carbon-oxygen nuggets quickly collapse until the repulsive forces between electrons begin to predominate over gravity, preventing any further collapse. The core is now in a semi-crystalline and metallic state—where the negatively charged electrons are free to swim through a solid matrix of positively charged nuclei. Astronomers call these compact stellar remnants *white dwarfs* because of their typically high surface temperatures (25,000 to 200,000 Kelvins) and incredibly small sizes (about as big as the Earth). Famous white dwarfs include Sirius B (first discovered in 1862 by the pioneering telescope builder Alvan Clark), the proto-white dwarfs at the centers of the Ring and Helix Nebulae, and the variable white dwarf ZZ Ceti, which is undergoing mini starquakes.[1]

With Sun-like masses collapsed into Earth-like volumes, white dwarfs have densities that greatly exceed anything that we can study in the laboratory. Consider a typical density of 10^6 grams/cm^3. A vial of the stuff would weigh several tons—the equivalent of a Toyota in a teaspoon. Further insights on white dwarfs have come from theoretical studies of how matter should behave at such high densities. The famous Indian astrophysicist Subrahmanyan Chandrasekhar was the first to reveal the bizarre traits of these collapsed objects. In 1930, as a first-year graduate student on a steamer voyaging from his home in India to Cambridge University in England, Chandrasekhar derived the equation of state that best describes white dwarfs. He found that these objects no longer obey the *Ideal Gas Law* where the pressure P, density of particles n, and temperature T are mutually dependent according to $P = nkT$. Instead, the pressure becomes solely dependent on the density—or equivalently the degree of crowding between particles. This so-called degenerate state of matter follows the Pauli Exclusion Principal of quantum mechanics, whereby no two particles can share the same exact quantum state. One can liken this situation to a hotel that has overbooked its rooms. Once the cheap rooms on the lower floors are occupied, all other guests would have to stay in the more expensive suites on the upper floors—even if they had reserved the least expensive rooms. So too, electrons in a degenerate state will attain higher energies and corresponding pressures according to how crowded they are. Temperature no longer plays any role in pressurizing the remnant.

Chandrasekhar determined that this peculiar state led to some rather bizarre behavior. Let's say you added lots of matter to the Sun. Its interior pressures and temperatures would both increase in response to the added weight. And that would likely cause the Sun to enlarge. We see just this sort of behavior along the main sequence, where stars with luminosities and corresponding masses greater than that of the Sun are also significantly larger than the Sun (see figure 5.2). For a white dwarf, however, any addition of matter will actually shrink the remnant, as it will be unable to heat up and enhance the thermal pressure in response to the overlying weight. There is a limit to this behavior, of course, when the remnant shrinks to zero radius. Chandrasekhar determined that this practical limit occurred at a mass of 1.4 Suns. Stellar remnants with higher masses must become something else. Thanks to advances at the crossroads of quantum and gravitational physics, we now have a pretty good idea what that "something else" must be.

Fates of Massive Stars

Stars with masses that exceed 8 Suns are able to fuse carbon into ever heavier elements—ultimately producing dormant iron cores. The masses of these cores are expected to exceed the Chandrasekhar limit of 1.4 Suns, where the pressure provided by electron degeneracy is not enough to keep the iron core from collapsing under its own weight. The resulting implosion takes less then a second. Once again, the end state depends on the mass of the core.

If the core mass is less than about 3 Suns, the core will become a neutron star. Above 3 or so solar masses, the core is expected to become a black hole. In both cases, the drastic infall of matter releases tremendous amounts of gravitational energy. We know about these occurrences, because we have witnessed supernova explosions whose awesome power cannot be explained in any other way. We can also see the aftermath of these titanic events in the form of supernova remnants.

Last Gasps

Massive stars rip through their lives so fast that it becomes difficult to discriminate between their "normal" and more "pathological" evolutionary phases. Like Wolfgang Mozart, Charlie Parker, Jimi Hendrix, Janis Joplin, John Belushi, and other *enfants terribles,* they illuminate with extraordinary brilliance, yet doom themselves in the process. We can see this unfolding drama in images of Wolf-Rayet (WR) stars and Luminous Blue Variable (LBV) stars, whose outbursts have produced circumstellar shells and bubbles of fluorescing gas (see figure 8.3). Spectroscopy of these sources also reveals something special—as the spectral absorption lines are often strongly blue-shifted relative to the same lines in emission. Astron-

---

FIGURE 8.3. Death rattles of massive stars, though fleeting, can be found in the star-forming disk of our Galaxy. (*Top left*) The two-lobed Homunculus Nebula is the result of an eruption by the Eta Carinae binary star system. The original outburst was observed in 1841. The Homunculus is located inside the much larger Carina Nebula—a naked-eye object of the southern sky. At an estimated distance of 7,500 light-years, the bipolar outflow currently spans about a half light-year. (*Top right*) Thor's Helmet (NGC 2359) is a fluorescing bubble of gas that was blown by the Wolf-Rayet star HD 56925 into its surrounding molecular cloud. It is located about 15,000 light-years away in the direction of Canis Major and measures an impressive 30 light-years in diameter. (*Bottom*) The blue supergiant star Sher 25 has recently ejected an equatorial ring and two polar caps of glowing gas. It resides inside NGC 3603—one of the richest and most powerful star-forming regions in the Galaxy. The NGC 3603 complex is located in Carina at a distance of about 20,000 light-years. It measures roughly 20 light-years across. [(*Top left*) N. Smith, J. A. Morse (University of Colorado) et al., Hubble Space Telescope, NASA and the European Space Agency. (*Top right*) Star Shadows Remote Observatory, National Optical Astronomy Observatories

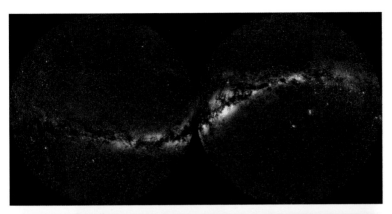

PLATE 1. Maps of the entire night sky can be made by taking deep images from both the northern and southern hemispheres of Earth and by then combining these images into large-scale mosaics. The appearance of these all-sky maps depends on what kind of two-dimensional projection is used. The top projection has the North Celestial Pole in the center of the left image and the South Celestial Pole in the center of the right image, with the celestial equator running around the perimeter of each image. The bottom projection has the galactic equator running across the middle, with the galactic center placed in the very middle of the map. Because this sort of projection is a more natural fit to studies of the Milky Way Galaxy, it is adopted in many of the images contained in this book. [Courtesy of A. Mellinger (Central Michigan University), whose renderings are available at http://home.arcor-online.de/axel.mellinger/mwpan_polar2 .html for the top image and http://home.arcor-online.de/axel.mellinger/ mwpan_aitoff.html, for the bottom image]

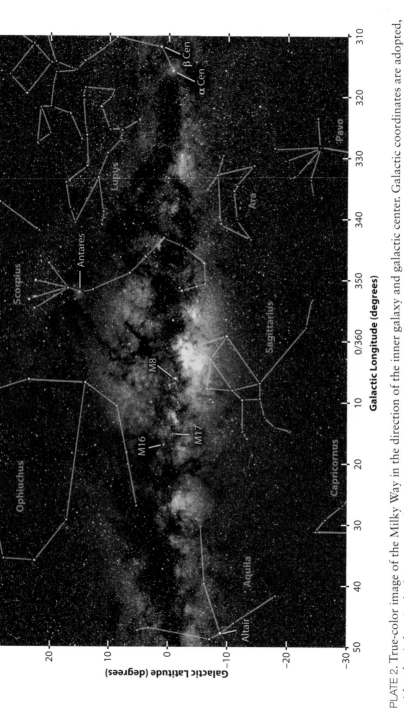

PLATE 2. True-color image of the Milky Way in the direction of the inner galaxy and galactic center. Galactic coordinates are adopted, with galactic longitude (l) increasing from right to left from 310° (−50°) through 0° (the galactic center) to +50°. Galactic latitude (b) runs from −30° through 0° (the galactic midplane) to +30°. The Rho Ophiuchus Nebula (l = 354°, b = +16°), Lagoon Nebula (M8) (l = 6°, b = −1°), and Eagle Nebula (M16) (l = 17°, b = +1°), are well-known telescopic objects within this range of galactic longitudes. These southern objects are best viewed from the southern hemisphere during the months of May–July. [Plates 2–5: Courtesy of A. Mellinger (Central Michigan University). Annotations by D. Lampert]

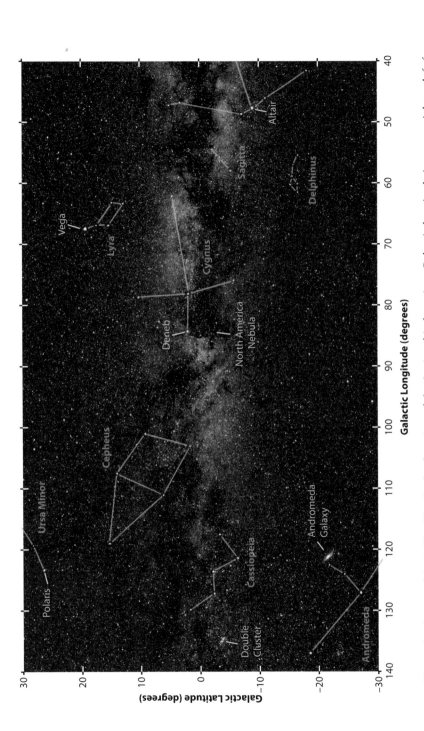

PLATE 3. True-color image of the Milky Way in the direction of the Sun's orbital motion. Galactic longitude increases right to left from 40° through 90° to 140°. The Cygnus Loop (l = 74°, b = −9°) and North America Nebula (l = 85°, b = −1°) can be observed with wide-field telescopes or large-aperture binoculars. They are best viewed through nebular filters that increase their contrast against the sky background. The Andromeda Galaxy is visible at (l = 121°, b = −22°). This part of the sky is best situated for nighttime viewing from northern latitudes during the months of August–October.

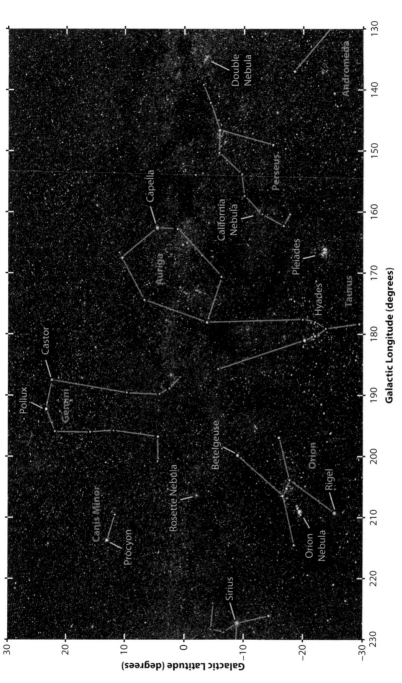

PLATE 4. True-color image of the Milky Way in the direction of the outer galaxy and galactic anticenter. Galactic longitude increases right to left from 130° through 180° (the galactic anticenter) to 230°. The Pleiades star cluster (l = 166°, b = −23°), Orion Nebula (l = 209°, b = −19°), and Rosette Nebula (l = 206°, b = −2°) are popular binocular and telescopic objects in this part of the sky. The months of November–January are most favorable for nighttime viewing of these objects.

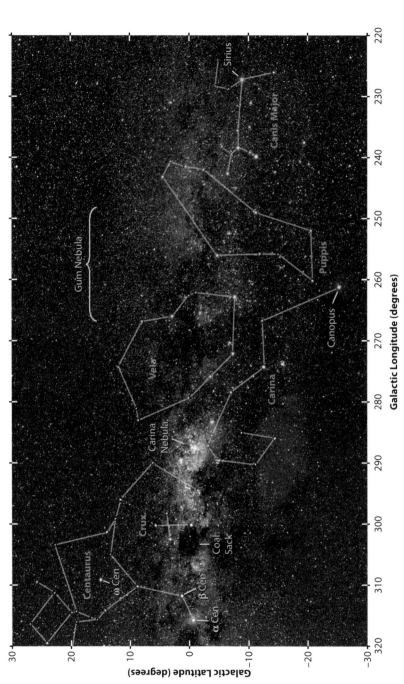

PLATE 5. True-color image of the Milky Way in the direction opposite to the Sun's orbital motion, Galactic longitude increases right to left from 220° through 270° to 320°. The Carina Nebula (l = 287°, b= −1°), Southern Cross (l = 300°, b = +3°), and Coal Sack (l = 302°, b = 0°) are naked-eye objects for observers of the southern sky during the months of February–April.

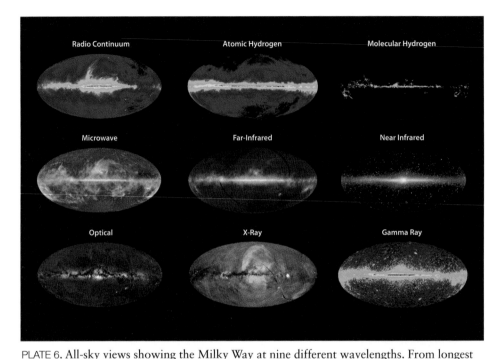

PLATE 6. All-sky views showing the Milky Way at nine different wavelengths. From longest to shortest wavelength (and correspondingly lowest to highest photon energy), they depict sources of radio continuum emission from hot ionized gas, radio line emission from warm atomic hydrogen, microwave line emission from cold carbon monoxide gas (which traces the abundant but much fainter molecular hydrogen gas), microwave continuum emission from cold dust (and the cosmic microwave background), far-infrared emission from star-warmed dust, near-infrared emission from stars, visible (optical) emission from stars and ionized gas, X-ray emission from compact stellar remnants and hot gas, and gamma-ray emission from the most energetic compact stellar remnants and diffuse gaseous outflows. [Compiled by D. Leisawitz (National Space Science Data Center–Goddard Space Flight Center–NASA), with the exception of the middle image. Radio Continuum (Bonn, Jodrell Bank, and Parkes Radio Observatories), Atomic Hydrogen (J. Dickey and J. Lockman/National Radio Astronomy Observatory), Molecular Hydrogen (Columbia University–Goddard Institute for Space Studies–NASA), Infrared (Infrared Astronomy Satellite–Infrared Processing and Analysis Center–Jet Propulsion Laboratory–Caltech–NASA), Near-Infrared (Cosmic Background Explorer–NASA), Optical (A. Mellinger), X-ray (Position Sensitive Proportional Counter–Roentgen Satellite), gamma ray (Energetic Gamma Ray Experiment Telescope–Compton Gamma Ray Observatory–NASA). Microwave image (*middle*) from the Planck mission, High-frequency Instrument and Low-frequency Instrument Consortia, European Space Agency, NASA]

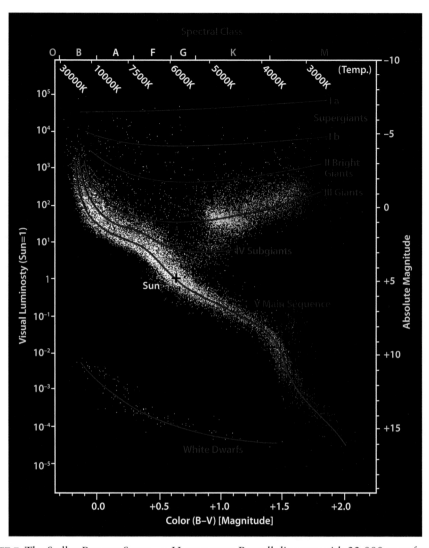

PLATE 7. The Stellar Rosetta Stone—a Hertzsprung-Russell diagram with 22,000 stars from the Hipparcos Catalogue and 1,000 low-luminosity stars (red and white dwarfs) from the Gliese Catalogue of Nearby Stars. All of the plotted stars have well-established distances and hence reliably determined luminosities. The "Color (B-V)" refers to a comparison of brightness measurements made through blue (B) and yellow (V) filters. This color index provides a quantitative measure of stellar surface temperature. The "Spectral Class" is based on the pattern of absorption lines evident in the star's spectrum. The resulting OBAFGKM sequence of spectral types also provides a good proxy for surface temperature. The "Absolute Magnitude" is a logarithmically compressed representation of luminosity. The Sun has an absolute visual magnitude of about +4.85. Families of stars include the luminosity-class V "main sequence," luminosity-class III giants, luminosity-class I supergiants, and the hot but dim white dwarfs. [Adapted from R. Powell, http://www.atlasoftheuniverse.com/hr.html, licensed under a Creative Commons Attribution-ShareAlike 2.5 License]

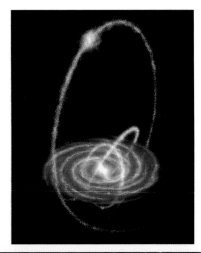

PLATE 8. (*Top*) Artist's depiction of the various stellar streams that are swarming through the galactic halo. Several of these streams have been identified with tidal tails from disintegrating globular clusters and dwarf galaxies. (*Bottom*) The globular cluster M15 dwells in the Milky Way's halo, more than 35,000 light-years from the Sun in the direction of Pegasus—the winged horse. M15 contains about a hundred thousand stars crowded into a region 120 light-years across. Most of the prominent stars in this ancient realm are red giants. The blue stars are not massive and young, but instead old giant stars that have lost their outer layers via winds or tidal stripping, thus exposing their hot blue interiors. The prominent blue source to the lower left is actually a planetary nebula—the gaseous remnant of such a star caught in the act of expiring. [(*Top*) R. Hurt, Spitzer Space Telescope, Caltech, Jet Propulsion Laboratory, NASA. (*Bottom*) Hubble Space Telescope, NASA, and the European Space Agency]

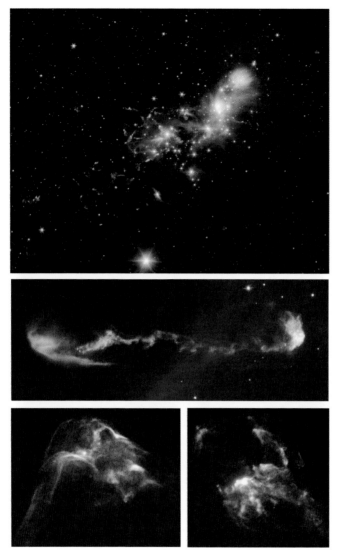

PLATE 9. (*Top*) The star-forming region NGC 1333 in the northern constellation of Perseus at infrared wavelengths, showing (in green) multiple ionized jets (Herbig-Haro Objects) from Young Stellar Objects. (*Bottom*) Visible-light views of the Herbig-Haro Objects HH47 (*upper*) which is located in the southern constellation of Vela and HH34 (*lower left*) and HH2 (*lower right*), both of which are in the equatorial constellation of Orion. HH47 shows a complex jet that has collided with gas and dust, producing shocked pileups near its hidden stellar source on the left and at its rightmost extremity. HH34 and HH2 show multiple bow shocks from jets interacting with the ambient gas and dust. [(*Top*) Spitzer Space Telescope (Infrared Astronomy Camera—IRAC). (*Bottom*) P. Hartigan (Rice University), NASA, and the European Space Agency]

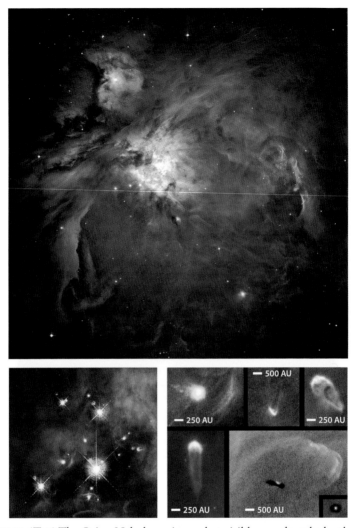

PLATE 10. (*Top*) The Orion Nebula, as imaged at visible wavelengths by the Hubble Space Telescope. This fluorescing tableau spans about a half-degree which, at a distance of 1,500 light-years, corresponds to dimensions of about 13 light-years on a side. (*Bottom*) Young stellar objects in the Orion Nebula include the "Trapezium" of hot O- and B-type stars along with several cometary nebulae being photo-evaporated by the most luminous Trapezium star Theta 1 Orionis C (*left*), along with proto-planetary disks ("proplyds") seen in silhouette against the nebular emission (*right*). The size scales are in astronomical units (AU), where 1 AU is the distance between the Sun and Earth. [(*Top*) M. Robberto and the Hubble Space Telescope Orion Treasury Project Team, Space Telescope Science Institute, NASA, and the European Space Agency. (*Bottom*) John Bally (University of Colorado), Dave Devine and Ralph Sutherland (Canadian Institute for Theoretical Astrophysics), NASA, and the European Space Agency]

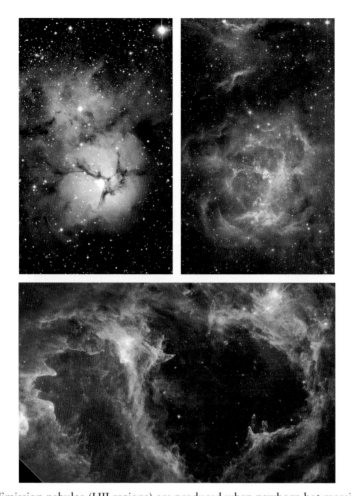

PLATE 11. Emission nebulae (HII regions) are produced when newborn hot massive stars ionize their placental surroundings. They represent some of the most dazzling objects in the Milky Way. (*Top*) The Trifid Nebula (M20) in Sagittarius as imaged in visible light (*top left*) and infrared light (*top right*). The visible light image features fluorescing ionized hydrogen gas in red and dust grains that are preferentially scattering blue starlight. The infrared image reveals the presence of organic molecules (coded in green) and warm dust grains (coded in red) that have been irradiated by the hot newborn stars. Image dimensions are roughly 27 × 44 light-years at an estimated distance of 5,400 light-years. (*Bottom*) The Soul Nebula (W5) in Cassiopeia dwarfs the Trifid and Orion nebulae by factors of 10 and 20, respectively. This infrared image reveals two large cavities in the nebulosity that have been excavated by the intense UV radiation and winds from hot O-type stars. Inward pointing pillars of exceptionally dense gas and dust have resisted the erosive processes. New generations of stars can be found in these compressed pillars. [(*Top left*) National Optical Astronomy Observatories, National Science Foundation. (*Top right*) J. Rho, Spitzer Science Center, Jet Propulsion Laboratory, Caltech, NASA. (*Bottom*) L. Allen and X. Koenig (Harvard-Smithsonian Center for Astrophysics), Jet Propulsion Laboratory, Caltech, NASA]

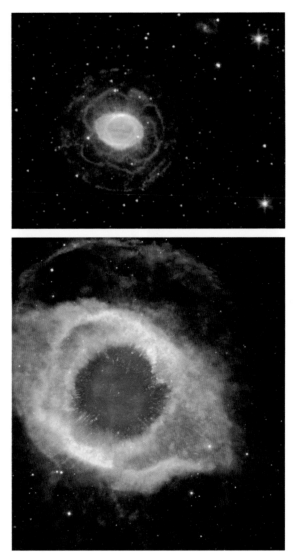

PLATE 12. Images of planetary nebulae at visible and infrared wavelengths attest to the complex architecture of these stellar exhalations. (*Top*) Mid-infrared view of the Ring Nebula (M57) in Lyra reveals an inner cylindrical shell of ionized gases a light-year in extent surrounded by several crenellated shells of cooler molecular gas that were ejected by windy outbursts from the rapidly evolving star at the center. (*Bottom*) A composite visible + infrared view of the Helix Nebula (NGC 7293) in Aquarius shows ionized gas (coded blue and green) mixed with molecular gases (coded red), with cometary streamers directed away from the central white dwarf. The nebula spans about 4 light-years. [(*Top*) J. Hora (Harvard-Smithsonian Center for Astrophysics), Spitzer Science Center, Jet Propulsion Laboratory, Caltech, NASA. (*Bottom*) J. Hora (Harvard-Smithsonian Center for Astrophysics), C. R. O'Dell (Vanderbilt University), Jet Propulsion Laboratory, Caltech, NASA]

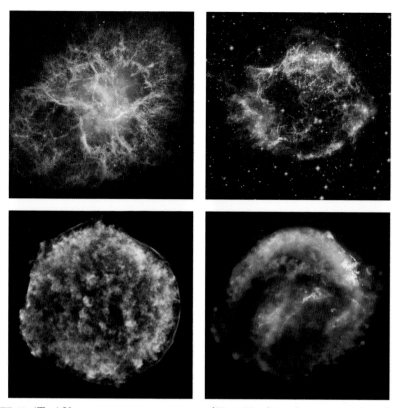

PLATE 13. (*Top*) Young supernova remnants of Type II, where the emission directly traces material from a massive star that exploded less than a thousand years prior. In these composite images, blue represents X-ray emission from million-degree gas, green denotes visible emission from thousand-degree gas, and red indicates infrared emission from cooler dust grains. (*Top left*) The Crab Nebula (M1) in Taurus as seen 953 years after the progenitor star exploded. The combined nebulosity spans about 12 light-years. The remnant neutron star appears as an X-ray bright pulsar near the center. (*Top right*) Cassiopeia A as seen 325 years after the initial supernova. The nebular remnant spans about 15 light-years. The hot compact remnant is the blue dot just below center. (*Bottom*) Supernovae explosions of Type Ia result from the obliteration of a white dwarf in a closely interacting binary star system. Historic supernovae of this type and their nebular remnants include (*bottom left*) Tycho's Supernova of 1572 (shown in X-rays with the higher-energy X-ray emission in blue), and (*bottom right*) Kepler's Supernova of 1604 (shown as a composite of X-ray emission in blue and green, visible light in yellow, and infrared light in red). [(*Top left*) X-ray: J. Hester (Arizona State University) et al., Chandra X-ray Center, NASA; Optical: J. Hester and A. Loll (Arizona State University), NASA + European Space Agency; Infrared: R. Gehrz (University of Minnesota), Jet Propulsion Laboratory, Caltech, NASA. (*Top right*) O. Krause (Steward Observatory) et al., Spitzer Science Center, Jet Propulsion Laboratory, Caltech, NASA. (*Bottom left*) J. Warren and J. Hughes (Rutgers University), Chandra X-ray Center, NASA. (*Bottom right*) R. Sankrit and W. Blair (Johns Hopkins University), NASA, and the European Space Agency]

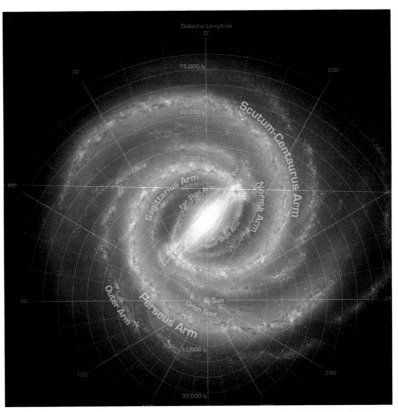

PLATE 14. A recent rendering of the stellar bar and spiral arms in the Milky Way, based on infrared observations with the Spitzer Space Telescope. The bar is tilted by about 45 degrees with respect to our sightline to the galactic center. The spiral structure is dominated by two main arms that emerge from the ends of the central bar. [R. Hurt (Spitzer Science Center), Jet Propulsion Laboratory, Caltech, NASA, based on data from the Spitzer Galactic Legacy Infrared Mid-Plane Survey Extraordinaire (GLIMPSE)]

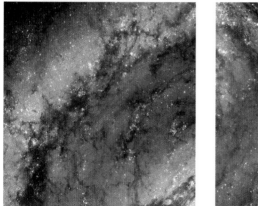 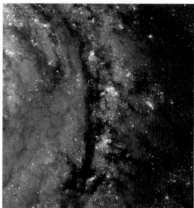

PLATE 15. (*Top*) The Whirlpool Galaxy (M51 = NGC 5194) is a well-known "grand-design" spiral galaxy, where each spiral arm is thought to be tracing the crest of a propagating density wave in the disk. Gravitational effects from the companion galaxy NGC 5195 are likely powering these robust waves. (*Bottom*) Close-ups of the spiral arms reveal spatial sequences of dark dust lanes, red HII regions, and blue stellar populations. These sequences trace the transformation of dusty gas clouds into evolving clusters of stars—a spatio-temporal process that is predicted by spiral density wave theory. [S. Beckwith (Space Telescope Science Institute) and the Hubble Heritage Team (Space Telescope Science Institute–Associated Universities for Research in Astronomy), NASA, and the European Space Agency]

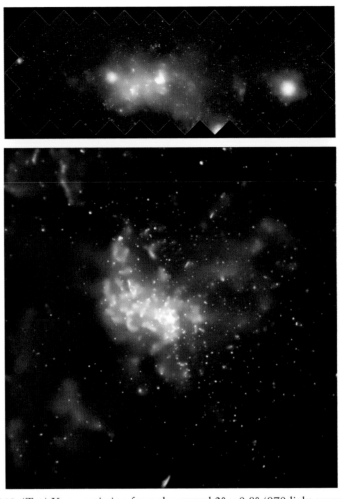

PLATE 16. (*Top*) X-ray emission from the central 2° × 0.8° (970 light-years × 390 light-years) in the Milky Way betrays high-energy processes associated with supernova remnants, the collective winds from newborn star clusters, and more than 20,000 point-like sources that likely involve compact stellar remnants in closely interacting binary systems. The X-ray emission has been color-coded according to its energy, with red representing X-ray photons of lowest energy and blue representing the highest-energy photons. The two bright sources on opposite sides of the galactic center are thought to be accreting neutron stars or black holes. (*Bottom*) Close-up of the X-ray emission from the central 8.5 × 8.5 arcmin (68 × 68 light-years) reveals more than 2,000 point sources along with diffuse emission from million-degree gas on multiple scales. This amazing image provides compelling evidence for powerful outflows on large scales and colliding winds from massive stars and pulsars on smaller scales. [(*Top*) D. Wang (University of Massachusetts) et al., Chandra X-ray Center, NASA. (*Bottom*) F. K. Baganoff (Massachusetts Institute of Technology) et al., Chandra X-ray Center, NASA]

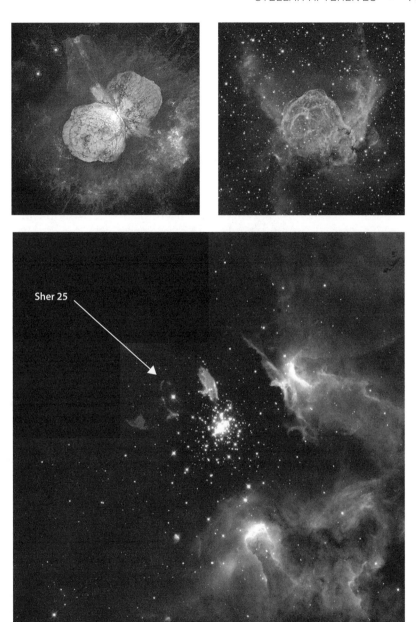

(NOAO), National Science Foundation (NSF). (*Bottom*) W. Brandner (Infrared Processing Center, Jet Propulsion Laboratory–Caltech), E. K. Grebel (University of Washington), Y.-H. Chu (University of Illinois Urbana–Champaign), Hubble Space Telescope, NASA, and the European Space Agency]

omers have interpreted this observed behavior as compelling evidence for stars that are literally blowing themselves away.

When a hot star blows a strong wind, it produces spectral-line emission from its extended atmosphere. It also produces absorption lines that are blue-shifted relative to the emission lines. That is because the absorbing material is blowing in our direction and so absorbs light at a Doppler-shifted wavelength that corresponds to the material's line-of-sight velocity. The combined profile of line emission and blue-shifted absorption is known as a "P-Cygni" profile, named after the B1 supergiant and Luminous Blue Variable star P-Cygni. Astronomers have used these spectroscopic profiles to clock the outflow velocities from massive hot stars. The prodigious winds are found to blow at thousands of kilometers per second, yielding mechanical "luminosities" that rival the radiant output from these most powerful stars. Such profligate behavior cannot be sustained for long—maybe a few hundred thousand years at best. During this time, the stellar blowhards are susceptible to profound instabilities and dramatic outbursts.

The LBV star Eta Carinae is worthy of special note, as it is one of the most unstable stars in the sky (see figure 8.3). First catalogued by Edmund Halley in 1677, it has since varied in dramatic and unpredictable ways. In 1843, it was seen to brighten by 6 magnitudes (a factor of 250), making it the second brightest star in the southern sky (after Sirius). It has since diminished back to its prior state, shining at fifth magnitude amidst the rest of the Carina Nebula complex. Along the way, it has inflated a bipolar nebula about a light-year in extent that appears to be throttled by a ring or disk of dense gas. Even in quiescence, Eta Carinae is incredibly powerful—with a visible light output equivalent to 4 million Suns. Its infrared luminosity is about a hundred times greater still, reminding us that most of its outpouring light is first absorbed by thick layers of dust. Continuous monitoring by the Hubble Space Telescope and various ground-based telescopes has led several astronomers to conclude that the star is actually an interacting system of two massive stars which come close enough to blast one another with their winds every 5.5 years. What most every astronomer agrees upon is the short-term nature of this system. Something very big is going to happen to Eta Carinae within the next 10 thousand to 100 thousand years.

Supernova Remnants

Once massive stars run out of ways to prop themselves up against gravity, their cores implode and their surrounding layers explode. The resulting pyrotechnics can last for up to a year or so. After the initial flash has ebbed, we can still trace the interstellar consequences via the nebular remnants that the explosions have created. For the first few thousand years, the supernova remnant is dominated by material that was ejected by the star itself.

The Crab Nebula (M1) is a prime example of this early stage. The supernova explosion itself was recorded to have occurred on July 4, 1054 CE, by Chinese astronomers who noted this event as a "guest star" next to the star currently known as Zeta Tauri.[2] Anasazi Indians of the American Southwest also saw something remarkable during this time. A rock painting in Chaco Canyon, in particular, shows a daytime "star" next to a crescent Moon— which would be consistent with both objects as they appeared in Taurus on that date. Today, long-exposure images show a diffuse S-shaped nebula riven by tendrils of line-emitting plasma (see plate 13). The diffuse nebulosity has a spectrum consistent with that of a synchrotron source, where electrons are twirling around lines of magnetic force at speeds close to that of light. The colorful tendrils bespeak torrid conditions where million-degree gas is condensing and cooling to "warm" line-emitting temperatures of "only" about 10,000 Kelvins. The hot gas itself can be seen in X-ray images of the nebula. Monitoring of the Crab Nebula by the HST has enabled astronomers to create a movie of its innermost activity (see http://chandra.harvard.edu/photo/2002/0052/anima tions.html). Here, the nebula can be seen responding to the energizing provocations of the remnant pulsar's wind. More on these compact oddities soon.

The Cassiopeia A radio source is another key example of a young supernova remnant (see plate 13). The brightest radio source in the sky beyond our Solar System, Cas A is also evident at infrared, visual, and X-ray energies. Spectroscopy of the line-emitting gases indicates expansion velocities of up to 10,000 kilometers per second. Such high speeds, when compared with the crosswise motions that have been observed, yield a distance of about 11,000 light-years, a size of about fifteen light-years, and a detonation date of 1681 CE—give or take twenty years. This comes tantalizingly close to an observation of a "new star" by Astronomer Royal John Flamsteed on August 16, 1680. Today, X-ray observations reveal a hot compact source near the nebula's center that could be either a neutron star or black hole that is accreting gas from its surroundings.

After a few thousand years, the supernova's shock waves will have traveled well beyond the exploding matter. These waves compress and excite the ambient interstellar medium, causing the available gases to glow in their own right. At first, the shocked gas expands and cools at rates that conserve the initial energy of the blast. But at later stages, the amount of plowed-up interstellar material begins to dominate the dynamics, so that momentum is conserved instead of energy. In this so-called snowplow phase, the slowing and cooling supernova remnant radiates away the bulk of the remaining energy. This stage can take tens to hundreds of thousands of years. Because they last a while, supernova remnants in their energy- and momentum-conserving phases are fairly abundant. Indeed, some of the prettiest supernova remnants that adorn the Milky Way are of this evolved ilk.

In the time that a volcano on Earth erodes to a nub, a supernova remnant will become a fossil of its former iridescent self. Perhaps some of the diffuse shells that characterize the far-infrared and radio maps of the Milky Way can be attributed to these supernova fossils (see figure 3.7). If so, they vivify the large-scale effects that supernovae can exert throughout the Milky Way. We will consider these more extended matters in the next chapter.

Neutron Stars and Pulsar Activity

Neutrons—subatomic particles with masses similar to protons but with zero charge—were first experimentally verified in 1932 by Sir James Chadwick. One way to think of a neutron is to imagine smashing a proton and electron together to produce a single particle of zero charge. Fast-forward one year from Chadwick's discovery of neutrons and we find the galactic astronomer Walter Baade and iconoclast scientist Fritz Zwicky hypothesizing the existence of stars made entirely of these neutral particles. By scrunching a dormant stellar core to unfathomable densities, the constituent atoms would literally implode—their electron clouds crammed into their nuclei until every nuclear proton was converted into a neutron. The implosion, in turn, could explain the extraordinary energies expended during supernova explosions. Though compelling, Baade and Zwicky's ideas laid as dormant as their neutral stellar cores until the discovery of pulsars.

By the 1960s, radio telescopes and receivers had become sufficiently sensitive to track the time signatures of celestial signals. Cambridge University astronomers Jocelyn Bell-Burnell and Anthony Hewish, working with a crude array of radio antennae, were the first to discover sources of rapidly varying radio emission. Beeping at fixed rates like finely tuned clocks, these radio sources were first dubbed LGMs after the fanciful notion that Little Green Men could be responsible for producing such remarkably regular transmissions. To avoid any semblance of bias, most radio astronomers settled on calling these sources "pulsars." Finding a natural solution to the pulsar phenomenon proved extremely problematic, however.

Stars of any conceivable size were incapable of pulsating at such high frequencies without ripping themselves apart. And even white dwarfs could not rotate fast enough to create the observed modulations. What was left was Baade and Zwicky's radical notion of extremely compact stellar remnants made entirely of nuclear matter and with rotation speeds close to that of light. Sporting magnetic fields as concentrated as their nuclear matter, these so-called neutron stars could whip up pencil beams of synchrotron emission that would sweep the sky with every rotation. If you happened to be caught in one of these sweeping beams, you would receive a lighthouse blast of radiation at regular intervals ranging from 10 seconds to a mere millisecond (1/1,000 of a second) depending on how fast the neutron star was spinning.

As more radio pulsars were discovered, the idea that they were highly magnetized and rapidly spinning neutron stars became increasingly favored. To this day, most astronomers agree that pulsars are most likely neutron stars whose magnetic field configurations produce pencil beams that just so happen to sweep across us. That means there are a lot more neutron stars out there which we cannot yet detect. A few hints of this larger population have been found by the Chandra X-ray Observatory, where compact sources of steady but extremely faint X-ray emission have been identified.

We are left with incredible stats—stellar remnants of 1.4 to 3 solar masses compressed to diameters of about 20 kilometers—the approximate size of a city or large volcano. A more Earth-bound analogy would be to shrink an aircraft carrier weighing several hundred thousand tons down to something no larger than the tip of a ballpoint pen. Corresponding densities of 100 trillion grams per cubic centimeter give new meaning to the nature of matter itself. If you released a teaspoon of neutron star on Earth, it would plummet through the Earth as if all that rock and magma were no more than air. The nuclear teaspoon would then blithely oscillate back and forth through the center of Earth to the opposite surface and back like a subterranean yo-yo.

Along with these extreme densities, we would encounter magnetic fields a billion times greater than that on the Earth or Sun, and surface gravitational forces strong enough to drastically alter the course of light. The importance of neutron stars within the context of the larger Milky Way is that they represent an important end state for stars of sufficiently high mass. The pulsar phenomenon notwithstanding, most neutron stars are undetectable. As such, they are a form of baryonic dark matter (MACHOS) that permeates the disk of the Galaxy along with faint subdwarf stars, white dwarfs, and black holes.

Black Holes

A dormant iron core whose mass significantly exceeds the equivalent of 3 Suns cannot be stopped from collapsing further. Even the degeneracy pressure exerted by nuclear matter is insufficient to withstand the crushing weight of such a core. Stellar astrophysicists reckon that stars of 40 or more solar masses would likely produce these sorts of fatally overweight cores. Not knowing of any greater resistive force in nature, they predict that the massive cores rapidly plummet into black holes of their own devise. The critical moment occurs when the core shrinks to a size where the velocity of escape from the core's surface equals that of light. That defining radius is known as the Schwarzschild radius after the German astrophysicist Karl Schwarzschild, who worked out the solution for a nonrotating black hole shortly after Albert Einstein published his general theory of relativity. For those of us who are not well-versed in the complex mathematics of general relativity, the Schwarzschild radius can be readily derived by equating the Newtonian gravitational and kinetic energies, yielding

$$R_S = 2 \ G \ M/c^2$$

where R_S is the Schwarzschild radius, G is Newton's constant of universal gravitation, M is the mass of the gravitating object, and c is the speed of light. For a stellar core of 5 solar masses, the corresponding Schwarzschild radius amounts to no more than 15 kilometers—the equivalent of a good-sized city. Once the core shrinks past this limit, it literally disappears from view. Any hapless photons that the object may emit are doomed to eternal entrapment.

According to the above formulation, any mass has a Schwarzschild radius. For example, the Sun has a Schwarzschild radius of about 3 kilometers, and the Earth's Schwarzschild radius is 300,000 times smaller—about 1 centimeter. The trick to making a black hole is to somehow crunch that mass into a volume that fits inside the Schwarzschild radius. That could never happen with the Earth or Sun, as the molecular bonding within the Earth and the pressure of degenerate electrons in what will become the Sun's white-dwarf remnant are sufficient to keep these bodies stable against further collapse. The dormant iron core of a massive supergiant star is another story altogether. We are left to speculate on the future of all that matter once it falls past the so-called event horizon of no return.

It is well beyond the scope of this book (or the expertise of its author!) to prognosticate upon what it might be like inside the event horizon of a stellar black hole. If you fell past the event horizon, would you truly disappear from the space-time that characterizes the universe as we know it? Would you ultimately find a singular point of infinite density? Would you emerge somewhere else in some other time? Suffice to say that you couldn't call home to report your findings. Pertinent to this book, however, is what remains within our particular space-time and how a stellar black hole behaves within the larger context of the Milky Way. This much is fairly certain. All that we could ever perceive would be the black hole's mass, charge, and angular momentum. The mass can be inferred for any black hole that happens to be in a binary system, where the companion star's motion can be tracked via the Doppler-shifting of the star's spectrum, and the mutual masses derived from these motions. Any gravitating dark body that exceeds 3–5 solar masses is most likely a black hole. To date, more than a dozen stellar black hole candidates in the Milky Way have been identified by this method.

A black hole's excess charge would be next to impossible to detect, as nature abhors unbalanced charges. Lightning bolts on Earth are just one example of nature rapidly redressing an imbalance of charges. A black hole's angular momentum may be detectable, however, as any rotation of the black hole would likely induce magnetic fields that could possibly extend beyond the black hole's event horizon. It is indeed possible that we may one day detect magnetized jets spewing from the immediate vicinity of a rapidly rotating black hole, or perhaps even pulsar activity hosted by such a black hole.

Other ways of detecting black holes arise from more violent circumstances, as will be considered next.

Explosions in the Night

Take any two stars in a close binary system and let them evolve. The more massive member will run through its main-sequence and

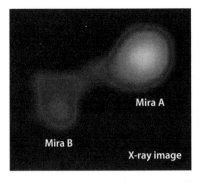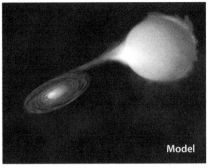

FIGURE 8.4. (*Top*) Cataclysmic variables are thought to involve closely interacting binary star systems, where the intense gravity of a compact remnant draws off material from the outer atmosphere of a nearby giant star. A model of this sort of interaction is shown on the right. Chandra X-ray observations of the Mira binary system appear to vindicate the model. As shown on the left, the white dwarf (Mira B) has distorted the atmosphere of the well-known red giant (Mira A). The ensuing interactions have caused both stars to glow in X-rays. [(*Left*) M. Karovska (Smithsonian Astrophysical Observatory) et al., Chandra X-ray Center, NASA. (*Right*) Illustration by M. Weiss (Chandra X-ray Center), NASA]

giant phases more quickly—leaving behind a white dwarf, neutron star, or black hole remnant. The less ponderous member will take longer to evolve—attaining giant status after its companion has expired and diminished into a compact remnant. If the giant and remnant are close enough, the compact remnant's gravity can siphon off material from the giant's bloated outer atmosphere. These intimate relations can lead to all sorts of fireworks in the Milky Way, several thousand of which have been recorded since the beginnings of human history. As a class, the so-called *cataclysmic variables* are all thought to involve stars and compact stellar remnants in closely interacting binary systems (see figure 8.4). There is an incredible variety within the genre, however. Given the unique nature of close relationships among humans, animals, and even plants, perhaps this is to be expected among these stellar couplings.

The most common outbursts derive from close binary systems, where white dwarfs are receiving material from their recently enlarged giant companions. If the accretion flow rate is high, the white dwarf exclaims its annoyance in the form of frequent but relatively low-luminosity outbursts. These are known as dwarf or recurrent novae. Although the outbursts are far from periodic, they typically occur every few months or so somewhere in the Galaxy for the dwarf novae and every few decades for the recurrent novae. The brightenings are thought to arise from the surrounding accretion disks, whose instabilities can lead to sudden flare-ups of hydrogen fusion.

Reduce the flow rate, and the white dwarf remains relatively docile until it can't take it anymore. Suddenly, the accumulated hydrogen on the dwarf's carbon-oxygen crust becomes completely involved in a thermonuclear runaway of chain reactions that tears across the surface like wildfire. The compact remnant rapidly brightens by factors of 10 thousand to 16 million (10–18 magnitudes) producing a new star-like source in the visible firmament. The classical novae that have been noted since before the time of Christ are today understood as arising from these sorts of close binary interactions. Each year, a few such novae are discovered in the Milky Way. Some of them have left nebular remnants—including the recurrent nova T Pyxidis (1966, 1944, 1920, 1902, 1890), Nova Persei 1902, and Nova Cygni 1992. Because the nova outburst does not destroy the white dwarf, encore performances are likely. You may have to wait 10,000 to 100,000 years to catch the next show, however.

If the mass of the white dwarf and the flow rate are "just right," the white dwarf can be taken completely unawares. Before its surface has had a chance to react, the white dwarf has accreted enough hydrogen to exceed the Chandrasekhar limit of 1.4 solar masses. The result of this "violation" is complete annihilation of the white dwarf and a supernova explosion even more powerful than those produced by massive supergiant stars. Astronomers designate these sorts of explosions as supernovae of Type Ia; massive core-collapse supernovae are of Type II or Ib. The Type Ia supernovae can be discriminated from their massive counterparts because their spectra are notably lacking in emission lines of hydrogen. That is because the exploding white dwarf consists mostly of carbon and oxygen, whereas the imploding core of a massive star still has lots of hydrogen surrounding it. Famous Type Ia supernova explosions include Tycho's Supernova which was recorded by the Danish astronomer Tycho Brahe in 1572, and Kepler's Supernova which first became visible in October 1604 (see plate 13).

If the interacting white dwarf is replaced by a neutron star or black hole, even more exciting displays are possible. That is because of the intense surface gravities of these extraordinarily dense objects. In essence the same accretion phenomena occur, but at higher velocities and corresponding energies. Consequently, one of the best ways to find neutron stars and black holes is via the X-ray and gamma-ray emission which their accretion disks produce when in closely interacting binary systems.

One of the most famous of these systems is SS 433—a cozy combo involving a neutron star or black hole remnant and a "donor" supergiant star of A7 Ib spectral type. The system is located inside the W50 supernova remnant about 18,000 light-years away in the constellation of Aquila the Eagle. Spectroscopic observations showed twin sets of emission lines—one blue-shifted, the other red-shifted—that then cycled back and forth every 162 days. As the comedian Father Guido Sarducci quipped in 1979 on an episode of the television show *Saturday Night Live*, "It seems-a to be a-coming and a-going." Astrophysicists today interpret these spectroscopic antics as arising from hot jets of material that are both streaming from the central remnant and precessing around like a corkscrew. Several turns of the screw can be seen in radio images of the system (see figure 8.5).

Neutron stars and black holes also provide viable models for explaining gamma-ray bursts—the most powerful explosions to be found in any galaxy. Gamma-ray bursts (or GRBs) last only seconds to minutes, but they pack an incredible punch. For the sake of comparison, the peak luminosity of a Type II supernova is the equivalent of 200 million Suns. A Type Ia supernova dishes out twenty-five times more radiant power, or about 5 billion solar luminosities at its peak. GRBs are yet another 200,000 times more luminous, yielding a million billion (10^{15}) Suns worth of instant energy. Indeed, they are roughly 100,000 times brighter in gamma rays than their host galaxies are at all wavelengths.

To explain these sorts of extreme phenomena, astrophysicists imagine that drastic reconfigurations of gravitating matter would be necessary. One possibility is the collapse of a rapidly spinning

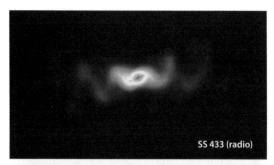

SS 433 (radio)

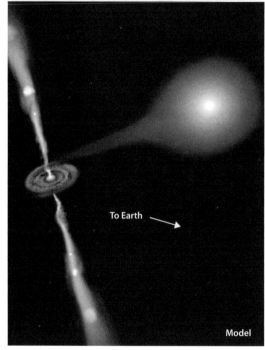

To Earth

Model

FIGURE 8.5. (*Top*) Radio image of the corkscrew stream that has been left behind by SS 433—a closely interacting binary system that includes an A-type supergiant and either a neutron star or black hole as its compact companion. (*Bottom*) Model of the system with its precessing jets. [(*Top*) K. Blundell and M. Bowler (University of Oxford), Very Large Array (VLA), National Radio Astronomy Observatory (NRAO), Associated Universities Incorporated (AUI), National Science Foundation (NSF). (*Bottom*) M. Weiss, C. Canizares, J. Kane, N. Schulz (Massachusetts Institute of Technology), Chandra X-ray Center, NASA]

supergiant star's core directly into a black hole—leading to a "hypernova." This so-called collapsar model appears most consistent with the enormous GRB that was recorded in the constellation of Leo on March 29, 2003. Follow-up observations of this GRB at X-ray and optical wavelengths confirmed that the exploding gases included elements that would have been cooked up in such a stellar conflagration.

Another option is to begin with a neutron star that accretes enough material—or collides with another neutron star—to transform into a black hole. The resulting release of gravitational energy would be sufficient to power the observed GRBs. The colliding neutron star model appears best able to explain the brief GRBs that have been recently discovered in elliptical galaxies, where star formation ended long ago. Bereft of short-lived massive stars, the elliptical galaxies could still play host to lingering neutron stars in binary death grips. To date, astrophysicists have yet to settle on a common explanation for the thousands of GRBs that have been recorded in the sky by orbiting gamma-ray instruments since the 1960s. However, the gravitational energies that are released when forging stellar black holes appear to be up to the task.

Galactic Theatrics

If you could simply turn a knob to compress time at will, you could witness a scintillating assortment of stellar outbursts as they light up the Milky Way like Times Square. These would include the many dwarf novae that occur each week to the few classic novae that are discovered each year, the supernovae that explode every few decades, and the gamma-ray bursts that are thought to bedazzle the Milky Way on 100,000-year timescales. Similar stellar outcries can be discerned well beyond the Milky Way—in the 100 billion other galaxies that populate the visible universe. Indeed, observations of extragalactic supernovae and GRBs have been vital to our understanding of these most powerful explosions. That is because they are much too infrequent within the Milky Way itself to be studied with any regularity (see figure 8.6).

For the supernovae of Type Ia, astronomers have exploited their nearly constant luminosities as "standard candles" for reckoning the distances of their host galaxies. From these distances and the corresponding redshifts of the galaxy spectra, they have learned that the universe of galaxies is expanding ever faster—as if propelled by some sort of invisible energy. The quest to determine the form of this *Dark Energy* has become one of the highest priorities of modern science.

As for the gamma-ray bursts, we now understand that virtually all of the observed GRBs exploded within some other galaxy in the

Gamma-ray Bursts

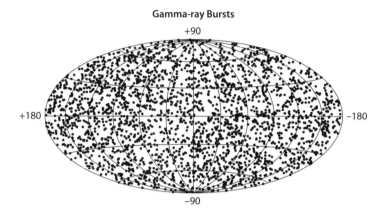

FIGURE 8.6. Spatial distribution of the 2,704 gamma-ray bursts (GRBs) that were observed across the sky by the Compton/BATSE experiment from 1991 to 1999. The lack of a preference toward the plane of the Milky Way and recent identifications of some GRBs with external galaxies at high redshift have indicated that most if not all of the observed GRBs originate from remote galaxies. [Burst and Transient Source Experiment (BATSE), Compton Gamma-ray Observatory, Space Sciences Laboratory, Marshall Space Flight Center, NASA]

far reaches of the universe. By observing the X-ray and optical afterglows of particular GRBs and measuring the redshifts of their spectral-line emission, astronomers have linked the outbursts to galaxies at distances of more than 10 billion light-years. These incredible distances have verified the truly spectacular nature of GRBs. We can only speculate on what it would be like if a GRB detonated within a few thousand light-years of the Solar System. Many of the consequences would depend on how narrow the GRB's beam of radiation was, and whether the Solar System happened to lie in the cross-hairs of that beam. Despite these uncertainties, several astronomers have concluded that the Earth's ozone layer would be fried—leading to all sorts of ecological emergencies in the Earth's biosphere. Further interactions between the gamma rays and the Earth's upper atmosphere could spawn a flood of radioactive nuclides that would spread across the planet and rain down on the surface. The likely consequence would be a mass extinction, similar to the major extinction events that have befallen Earth every 100 million years or so. Perhaps it is no coincidence that the geological record of mass extinctions approximates the estimated time between GRBs from sources within a few thousand light-years of Earth!

The above speculations have recently received corroborating support from observations of unusual outbursts within the Milky Way. One such event occurred December 27, 2004, when the highly magnetic stellar source known as SGR 1806-20 was observed to erupt in X-rays and gamma rays. This so-called *magnetar* is thought to reside 50,000 light-years away, clear on the opposite side of the Galaxy. Its observed behavior is consistent with that of a rapidly spinning neutron star with a magnetic field that is a thousand trillion (10^{15}) times greater than the average fields on Earth or the Sun. Every so often, such a magnetar will undergo a "starquake" which reconfigures its intense magnetic field and in so doing, unleashes prodigious salvos of high-energy radiation. In the case of SGR 1806-20, the flaring magnetar briefly outshone the active Sun as seen from Earth—despite its incredible distance. Results on Earth included a sudden ionospheric disturbance that oscillated in time with the magnetar's pulsating outburst. Imagine such a beast at a more modest distance of 5,000 light-years, and you've got the makings of a death star.

The flaring magnetars along with the extragalactic supernovae and GRBs demonstrate the importance of linking rare astrophysical phenomena within the Milky Way to more remote phenomena observed at greater frequency elsewhere in the universe. Our need to look for precedents beyond the Milky Way grows even greater in the next chapter, as we struggle to fathom the true form and function of our home galaxy.

CHAPTER 9

· · • • · ·

THE GALACTIC GARDEN

We are stardust—billion year-old carbon,
We are golden—caught in the devil's bargain,
And we've got to get ourselves back to the garden.
—Joni Mitchell in "Woodstock" (1969 CE)

IN THE COURSE OF THIS BOOK, we have encountered myriad stars of differing mass, color, size, luminosity, composition, age, and activity; clusters of stars in various configurations and stages of evolution; cold, dusty clouds pregnant with embryonic stellar and planetary systems; emission nebulae irradiated by hot young stellar upstarts; and other emission nebulae powered by bizarre stellar remnants—including white dwarfs, neutron stars, and black holes. Together, these variegated species inhabit the Milky Way, co-evolving and interacting as parts of a vast galactic ecosystem. The picture of a galactic garden (or jungle!) comes to mind, as it provides a helpful way to envisage the larger Milky Way in terms of its sundry contents.

Just as a garden's flowers, ferns, bushes, and trees can manifest multiple patterns on a variety of scales, so too the Milky Way's diverse stellar and nebular components reveal structures upon structures that bespeak a hierarchy of complex and interacting dynamics at work. Tracing out these structures has turned out to be a difficult challenge, alas. Our status as "insiders" within the Milky Way has pretty much prevented us from ever getting a clear view of the Milky Way as it would appear from the outside. We are therefore left to reconstructing such a view from our compromised impressions of the Milky Way along with observations of other galaxies that can serve as helpful prototypes.

Where to begin? One way is to start with our local circumstances and move out. This was partially attempted in chapter 4, where we encountered Gould's Belt of short-lived massive stars and the Local Bubble of hot gas. Larger structures were revealed in chapter 5, where the O- and B-type stars within 5,000 light-years of the Sun appeared arrayed in parallel bands that may be part of larger spiral arms. But these glimpses provide much less than the grand design that we aspire to perceive. To really make progress, it is necessary to "go long" and consider the full sweep of emission from the Galaxy.

Radio Reconnoitering

After the intrepid attempts by Herschel, Kapteyn, and Shapley to map the Galaxy's stellar structure (see chapter 2), it became clear that further progress would require sensing the Milky Way at nonoptical wavelengths—where the obscuring effects of dust are less pronounced. The first determined efforts to map the Galaxy's structure occurred at radio wavelengths. These began in the 1950s with single-dish surveys at crude angular resolutions of about a degree, and improved as interferometric arrays of radio dishes came online. One of the first comprehensive radio mappings of the Galaxy was made with single-dish telescopes in Holland and Australia. Tuning into the 21-centimeter emission line of atomic hydrogen, the radio surveyors obtained thousands of spectral scans, where the Doppler-shifted emission could be diagnosed in terms of line-of-sight velocities and positions within the Galaxy (see figure 9.1).[1]

This sort of mapping is tricky, however, as it relies on the measured radial velocities as proxies for distance. Inside an ideal disk galaxy with gas in perfect circular motion about the galactic center, it is possible to measure the radial velocity of some emitting gas at a particular longitude and thence infer the spatial location of that gaseous parcel. In the actual Milky Way, alas, the radial veloc-

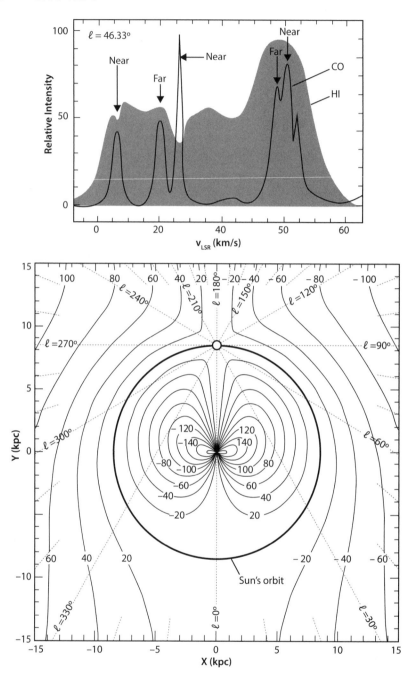

FIGURE 9.1. (*Top*) H I and CO spectra from emitting gas along a line of sight in the Milky Way corresponding to a galactic longitude of 46.33°. The H I emission comes from diffuse clouds of atomic hydrogen gas at many different line-

ity of gas at any particular longitude is a vexing index of the gaseous distribution in the Milky Way along with any peculiar velocities that result from the gravitational influences of stellar and gaseous accumulations in that direction. Such aberrant motions are well-known among the stars but are also present in the more quiescent gas, as it responds to the asymmetric influences of gravitating bars, spiral density waves, and galactic companions. Like the incongruity of the "tail wagging the dog," the gaseous motions in the Milky Way may be telling us more about the perturbative dynamics in the disk than the spatial distribution of the gas itself. Consequently, early mappings of the atomic hydrogen distribution turned out to be more an artifice of "streaming motions" along spiral arms and "velocity crowding" along tangents to the gaseous orbits than representations of any actual structure in the Milky Way (see figure 9.2).

Similar problems have been encountered in mappings of the molecular gas in the Milky Way. The CO-emitting gas yields more discrete mappings, as the cold molecular clouds are more concentrated than the diffuse atomic gas (see figure 9.1). Yet even with

---

of-sight velocities. The resulting spectrum shows a complex of Doppler-shifted emission features, each feature having a characteristic radial velocity relative to the "local standard of rest" (v_{LSR}—see Glossary). The CO emission comes from denser, more discrete molecular clouds which are easier to discriminate as separate line-emitting entities with distinctly different radial velocities. Near and far features refer to the "near-far distance ambiguity" interior to the Sun's orbit, where each radial velocity translates to both a "near" distance and "far" distance (see bottom). (*Bottom*) Template for converting observed radial velocities along longitudinal sightlines into positions within the Milky Way's disk. The Sun's position is denoted by the small circle near the top. Each plotted line denotes a contour of constant radial velocity. This mapping assumes perfectly circular orbits with velocities that vary only as a function of orbital radius. Unfortunately, neither of these assumptions is correct—making spatial maps of galactic structure based on radial velocities alone dicey at best.

[(*Top*) Adapted from spectra obtained as part of the Arecibo radio telescope survey of galactic HI emission and the Boston University—Five College Radio Astronomy Observatory Galactic Ring Survey of CO emission, National Science Foundation (NSF) (see http://www.ras.ucalgary.ca/VGPS/grs-arecibo-spec.gif). (*Bottom*) Adapted from map by Robert Benjamin (University of Wisconsin–Whitewater) (see http://wisp11.physics.wisc.edu/benjamin/pics/rotcurvemap.jpg)]

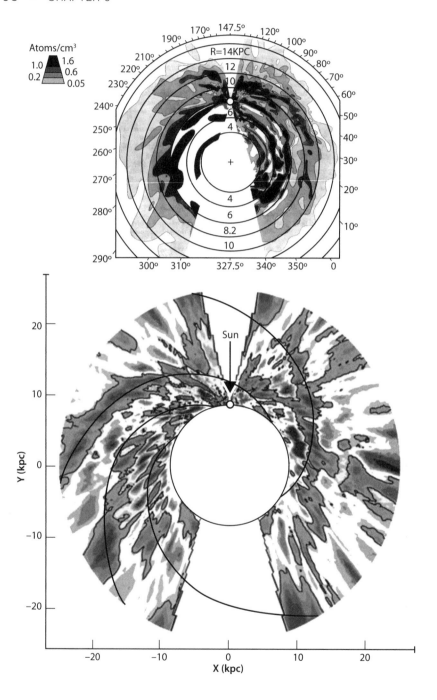

this advantage, the use of Doppler-shifted CO emission to fathom galactic structure has led to all sorts of uncertainties and controversies. Some progress has been made, with several radio astronomers claiming a "molecular ring" or tightly wound spiral pattern of CO-emitting gas at a galactocentric radius of 12,000 light-years—roughly half the distance to the solar orbit (see figure 3.5). Others have used kinematic mappings such as that in figure 9.3 to argue for chains of giant molecular clouds that trace out spiral arms in the Milky Way. Without more reliable measures of distance to these clouds, however, the overall renderings of gas in the Galaxy remain to this day stubbornly "nebulous."

One can do slightly better with observations of HII regions—those emission nebulae that have been ionized by newborn hot stars. If you can get a bead on any of the stars inside these regions, then it is possible to reckon a distance based on the star's spectral type and observed brightness. If an entire cluster of stars can be diagnosed in this way, the determination of distance becomes even more robust. Uncertainties relating to the amount of obscuring dust inevitably rear their hazy heads, but they can be beaten down to acceptable levels by observing the stellar colors, comparing them with expectations based on the stars' spectral types, and so deriving the reddening and overall extinction produced by the dust. Over the past three decades, several teams of astronomers

---

FIGURE 9.2. (*Top*) Rendering of the atomic hydrogen distribution in the Milky Way, based on early mappings of the positions and radial velocities of the radio-emitting gas. These sorts of mappings are especially sensitive to noncircular deviations in the gaseous orbits—leading to gaseous structures that are recognized today as being mostly spurious. Note that longitudes along the perimeter of the figure are offset from the current values by 32.5°. (*Bottom*) More recent mapping of the atomic hydrogen located beyond the Sun's orbit, where distance ambiguities are less severe. This rendering takes into account known noncircular motions and so should be more reliable. Candidate spiral arms are drawn. [(*Top*) Adapted from J. H. Oort, F. J. Kerr, and G. Westerhout, "The Galactic System as a Spiral Nebula," *Monthly Notices of the Royal Astronomical Society* 118 (1958): 379. (*Bottom*) Adapted from E. S. Levine, L. Blitz, and C. Heiles, "The Spiral Structure of the Outer Milky Way in Hydrogen," *Science* 312, no. 5781(2006), 1773–1777; courtesy of L. Blitz]

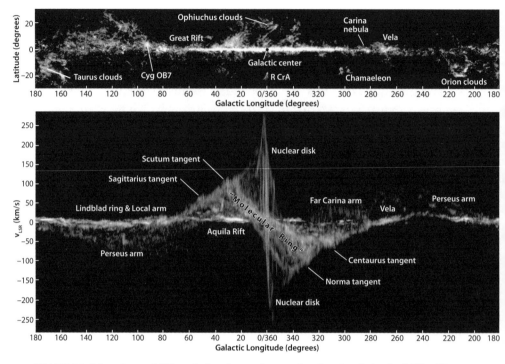

FIGURE 9.3. Mappings of CO emission from giant molecular clouds in the Milky Way. (*Top*) Standard mapping with the *x*-axis depicting the galactic longitude, and the *y*-axis corresponding to the galactic latitude. Prominent molecular clouds are noted. (*Bottom*) Kinematic mapping with the *x*-axis again marking the galactic longitude but with the *y*-axis indicating the radial velocity of the emitting gas. Special features in the galactic disk have been identified based on their unique pattern of longitudes and velocities. They include the nuclear disk with its extreme range of radial velocities, the molecular "ring" whose tilted pattern of velocities indicate fairly normal circular rotation, the Perseus and Carina spiral arms, and tangents to the Carina, Centaurus, Norma, Scutum, and Sagittarius arm segments. [Courtesy of Tom Dame with reference to T. Dame, W. Hartmann, and P. Thaddeus "Milky Way in Molecular Clouds: A New Complete CO Survey," *Astrophysical Journal* 547 (2001): 792]

have tried their hands at this sort of nebular fathoming. The results have been mixed, especially when the few HII regions containing observable stars are augmented with the more numerous HII regions that only have their radio hydrogen-line emission to observe and diagnose. In the latter case, one is again left to estimating distances based on the observed Doppler shifting of the spectral lines, the corresponding radial velocities of the emitting clouds, and their ultimate translation to distance based on some accepted rotation curve for the Galaxy. Alas, it appears that with increasing numbers of HII regions, the delineation of any spiral pattern grows increasingly muddled (see figure 9.4).

Perhaps surprisingly, the most often cited mapping of gaseous structure in the Milky Way comes from radio observations of pulsars. Radio signals from pulsars are perceptibly affected by scattering in the interstellar medium. The twinkling of starlight by scattering in the Earth's atmosphere provides a helpful analog to the scintillation that radio astronomers observe when monitoring pulsar signals. For the pulsar radio emission, the scattering is produced by free electrons in the interstellar medium. The charged electrons interact strongly with the electromagnetic radiation—absorbing, re-emitting, and so briefly delaying the radio waves according to the electrons' material concentrations and magnetically constrained motions. By observing the wavelength-dependent fluctuations in a pulsar's blinking signal (its so-called dispersion measure), astronomers can surmise the number of free electrons along the line of sight to that pulsar. And by monitoring enough pulsars at differing distances from us, they can infer the overall distribution of free electrons in the Galaxy. Their resulting map indicates a four-armed pattern of concentrated electrons, each with an average "pitch angle" (relative to circular rings) of about 20 degrees (see figure 9.4). If the electrons are tracing the ionized gas component in the Milky Way, then their distribution provides a handy means of tracing regions where massive stars are being formed, living out their torrid lives, and explosively expiring. These sorts of energized ecosystems are just what we would expect to occur within our Galaxy's spiral arms.

Stellar Spelunking

Like explorers of vast underwater caves, galactic astronomers must piece together limited sensory input in order to derive reasonable semblances of their surroundings. The obscuring effects of dust, in particular, have prevented optical astronomers from ascer-

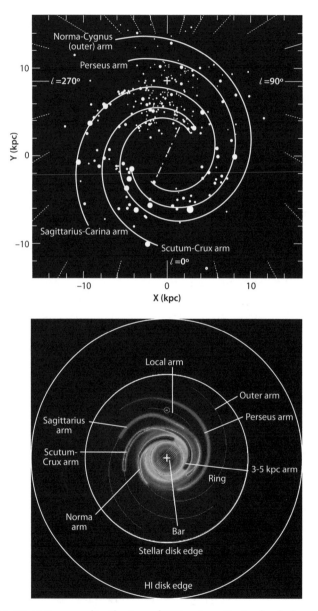

FIGURE 9.4. (*Top*) Face-on distribution of HII regions (ionized nebulae associated with newborn hot stars), with model spiral arms overlaid. Sizes of the circles are proportional to the HII region luminosities in the radio continuum. Distances to these nebulae are based on a mix of photometrically derived distances to the underlying stars and kinematically derived distances to the line-emitting gas. To some astronomers, the HII regions appear to delineate a four-armed spi-

taining the overall distribution of stars in the Milky Way. Our optical view is especially hindered toward the inner galaxy, where the abundance of dusty clouds has constrained our stellar spelunking to only a few thousand light-years from the Sun. These murky circumstances changed in the 1990s, when the Cosmic Background Explorer (COBE) satellite surveyed the entire sky at infrared and microwave wavelengths. The near-infrared maps at 1.25, 2.2, 3.5, and 4.9 micron wavelengths were especially important to the study of stellar structure in the Galaxy, as most of the stars in the Milky Way are low-temperature near-infrared emitters. Also, because infrared light is relatively unimpeded by intervening dust, the COBE near-infrared maps could reveal stars from throughout the disk (see plate 6). Close examination of these maps revealed a puzzling asymmetry in the central concentration of stars. This was interpreted as indicating the presence of a central bar that is pointed almost toward the Sun. The near part of the bar (seen at positive longitude) is closer to us and hence appears a bit larger than the farther part (seen at negative longitude). Careful analysis of these maps yielded a rendering of the Galaxy's stellar disk that features both the central bar and an outer warping of the disk (see figure 3.8).

Completed in 2001, the ground-based Two Micron All Sky Survey (2MASS) greatly improved on COBE by mapping the near-

ral pattern. In this rendering, the position of the Sun is marked with a cross. Compare with figure 5.10, which shows the local distribution of optical HII regions and star clusters containing young hot massive stars. (*Bottom*) Rendering of the free electron distribution in the Milky Way, based on the scintillating effects that these charged particles produce on pulsar signals. The free electrons are thought to provide a reliable tracer of gas that has been ionized by the star-forming activity in the Galaxy's spiral arms. [(*Top*) Adapted from D. Russeil, "Star-forming Complexes and the Spiral Structure of our Galaxy," *Astronomy & Astrophysics* 397 (2003): 133–146. (*Bottom*) Adapted from rendering by the Canadian Galactic Plane Survey team (see http://www.ras.ucalgary.ca/CGPS/where/plan/), courtesy R. Taylor (University of Calgary), based on findings of J. M. Cordes and T.J.W. Lazio (2003), "Using Radio Propagation Data to Construct a Galactic Distribution of Free Electrons, http://arxiv.org/abs/astro-ph/0301598, with further reference to J. H. Taylor and T.J.W. Cordes, *Astrophysical Journal* 411 (1993): 674]

infrared sky at 500-times finer spatial resolution (about 4 arcseconds). The resulting harvest of a half billion stars has provided astronomers with a rich catalog of stellar magnitudes, colors, and positions in the disk (see figure 9.5). By culling carbon-rich red giant stars according to their distinctly red colors, one astronomical team found about 30,000 of these ruby orbs delineating the central bar. The carbon stars have fairly constant luminosities equivalent to about 5,000 Suns each, and so could be used as "standard candles" for fathoming the distances to different parts of the bar. The resulting configuration has the bar tilted by 30 degrees with respect to the line of sight that connects the Sun with the galactic center. About 15,000 light-years long and 5,000 light-years wide (a 3:1 aspect ratio), the bar extends to more than 7,500 light-years from the galactic center (one-third the distance to the Sun), where it appears to merge with the "pseudo-ring" of molecular gas that has been found beyond this radius.

These results were confirmed in 2008, when a more sensitive near-infrared mapping of the Milky Way by the Spitzer Space Telescope revealed a similar configuration for the central bar. Even more exciting, members of the survey team found compelling evidence for a two-armed spiral pattern in the stellar disk. By tagging red horizontal-branch giant stars according to their particular near-infrared colors, the investigators perceived regions in the Milky Way that were especially rich in these stars, and so were likely tracing long sightlines through spiral arms. Though expecting star counts that would support the four-armed spiral pattern seen in the gas, they found that the Perseus and Scutum-Centaurus arms had stellar counterparts which connect nicely with the central bar, but the Sagittarius and Norma arms were for the most part missing (see plate 14).

This startling result may seem at first to be a contradiction. However, other spiral galaxies also show wavelength-dependent disparities in their spiral-arm patterns. Images at wavelengths that trace the gas, dust, and recent star formation often yield greater numbers of spiral arms and intermediary "spurs" than do images of the underlying near-infrared starlight (see figure 9.6). Here, we are likely witness to the interplay between density waves in the

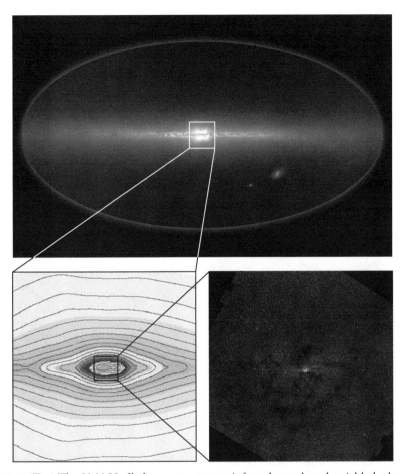

FIGURE 9.5. (*Top*) The 2MASS all-sky survey at near-infrared wavelengths yielded a bounty of 500 million stars. (*Bottom left*) Contoured star counts in the inner 12° × 12° (6000 ly × 6000 ly) reveal a greater extent in latitude at positive longitudes (toward the left). This has been interpreted as the consequence of viewing the near end of a stellar bar. (*Bottom right*) Close-up of the central 2° × 2° (1000 ly × 1000 ly) at near-infrared wavelengths is peppered with myriad stars. Analysis of these stars has yielded distances that indicate a distinct bar-like structure for the central "bulge." [(*Top*) Atlas image obtained as part of the Two Micron All Sky Survey (2MASS), a joint project of the University of Massachusetts and the Infrared Processing and Analysis Center/California Institute of Technology, funded by the National Aeronautics and Space Administration and the National Science Foundation. (*Bottom left*) Based on the findings of C. Alard, "Another Bar in the Bulge," *Astronomy & Astrophysics Letters* 379 (2002): L44, from analysis of 2MASS data (see http://www.obspm.fr/actual/nouvelle/nov01/alard.en.shtml). (*Bottom right*) Data from the Two Micron All Sky Survey (2MASS), University of Massachusetts, and the Infrared Processing and Analysis Center, NASA, and NSF. Data accessed by W. H. Waller from the SkyView interactive archive at NASA's Goddard Space Flight Center (see http://skyview.gsfc.nasa.gov)]

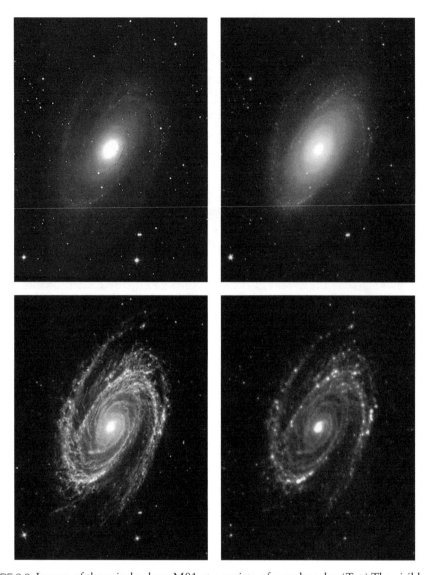

FIGURE 9.6. Images of the spiral galaxy M81 at a variety of wavelengths. (*Top*) The visible (*left*) and near-infrared 3.6-micron (*right*) views trace the distribution of stars in the disk. They show a relatively simple low-amplitude two-armed spiral pattern. (*Bottom*) The mid-infrared 8.0-micron view (*left*) traces the distribution of organic molecules, where they are being excited to glow by the presence of newborn massive stars. It shows a multiplicity of spiral arms and intermediary spurs. The 24-micron view (*right*) traces the dust complexes that have been warmed by these same stars. Multi-wavelength images of nearby spiral galaxies such as M81 provide important clues to understanding the stellar and gaseous architecture of our own Milky Way Galaxy. [(*Visible*) N. A. Sharp (National Optical Astronomy Observatories), Associated Universities for Research in Astronomy, National Science Foundation. (*Infrared*) K. Gordon (University of Arizona) and S. Willner (Harvard-Smithsonian Center for Astrophysics), Spitzer Science Center, Jet Propulsion Laboratory, Caltech, NASA]

stellar disk and the complex response to these waves as played out by the gas, dust, and consequent star-forming activity.

Given all these uncertainties and tantalizing clues, what can we conclude about the overall layout of our galactic garden? Perhaps most important is our new-found evidence for a substantial central bar in the stellar disk. This fundamental "overtone" in the rhythmically responsive disk is likely responsible for driving further oscillations throughout the stellar and gaseous components. Immediately surrounding the bar, two small arms of molecular gas have been traced by radio astronomers. The "3-kpc arm" on the near side of the nucleus was known back in the 1980s, but its far-side counterpart was only detected in 2008. Near the ends of the bar, some 7,500 light-years from the center, a ringlike configuration suggestive of two tightly wound spiral arms is evident in CO mappings of the molecular gas (see figure 3.5). This "pseudo-ring" is also aglow with radio emission from numerous HII regions—testament to its robust star-forming activity. Were we able to view the Milky Way from its brethren galaxies in the Local Group, we would find this inner annulus particularly dazzling. Beyond the stellar bar and gaseous pseudo-ring, we would see a simple two-armed pattern of enhanced stellar and gaseous density. And in between these primary arms, we would also perceive two additional arms and multiple spurs as traced by their gaseous contents and associated star-birth activity. Beyond all that, we could still trace a warping in the disk of stars and gas that inhabit these outer reaches.

Though other astronomers may disagree, I would recommend that the galaxy M109 (NGC 3992) provides the best nearby analog to what we have found in the Milky Way (see figure 9.7). This intermediate-class spiral galaxy is located about 46 million light-years away in the constellation of Ursa Major. Available optical images reveal a central bar, from which 2–4 spiral arms appear to unfold. The intermediate pitch angle of these arms is consistent with the 15°–20° pitch angles that have been surmised for our spiral arms. Then again, I'm not that certain about it. The Milky Way's molecular pseudo-ring interior to the Sun's orbit is likely hosting all sorts of burgeoning starburst activity and so should appear prominently at optical wavelengths. If so, the ringed-barred

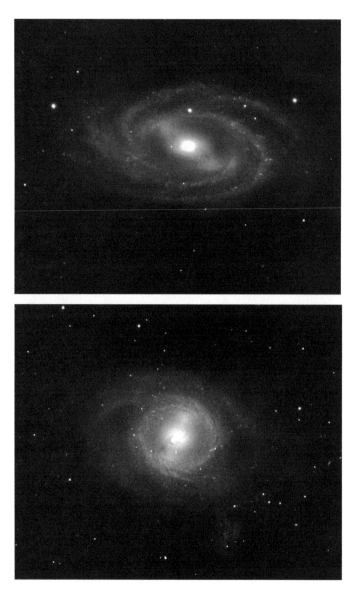

FIGURE 9.7. (*Top*) Intermediate-class spiral galaxy M109 (NGC 3992) has many of the structural attributes that are thought to characterize the disk of the Milky Way Galaxy. These include a central bar and 2–4 spiral arms that emanate from the two ends of the bar. (*Bottom*) The ringed-barred galaxy M95 (NGC 3351) may provide a closer analog to the particular situation in the Milky Way—interior to the Sun's orbit—where a ring (or a pair of tightly wound spiral arms) of molecular gas and star-forming activity is thought to be operating. [(*Top*) S. Swanson and Adam Block, Advanced Observers Program, National Optical Astronomy Observatories, Associated Universities for Research in Astronomy, National Science Foundation. (*Bottom*) M. McGuiggan and Adam Block, Advanced Observers Program, National Optical Astronomy Observatories, Associated Universities for Research in Astronomy, National Science Foundation]

galaxy M95 (NGC 3351) could provide a better prototype for our own galaxy (see figure 9.7). Located 38 million light-years away in the constellation of Leo, M95 sports two very tightly wound spiral arms that encircle its central bar. Other more loosely wound spiral arms can be seen beyond the pseudo-ring, as appear to be the case in our home galaxy.

Dynamics in the Disk

Although the true spatial structure of the Milky Way remains controversial, we can at least state with certainty that it includes a thin stellar and gaseous disk that is in differential rotation (see figure 9.8). Unlike a spinning compact disk, where every part is fixed with respect to every other part, the inner parts of our galactic disk are wheeling around at greater rates than the outer parts. This shearing motion would make any material spiral arm wind up over time until it's as tight as a watch spring. When we look at other spiral galaxies, however, the spiral arms are far looser affairs—with no more than two windings per arm. Clearly, spiral arms are somehow able to adorn their host galaxies without winding themselves into oblivion.

One solution to this so-called winding dilemma is to let the spiral arms be material but to make them transient phenomena—random enhancements in density and star-forming activity that get stretched out by the shearing motions into stubby spiral segments. After tens of mega-years, the massive stars and their luminous consequences peter out, leaving the remnant spiral segments to dim into obscurity well before they wind up (see figure 9.8). This sort of haphazard and ragged mode of inducing spiral structure may explain some of the less grandly designed spiral galaxies in the local universe. The galaxies NGC 4414 and NGC 5055 (M63) provide good examples of this genre (see figure 9.8).

If galaxies can be likened to vast gardens, then these sorts of spiral galaxies would most closely resemble the effusive and free-flowing gardens of English design.

Another key way to explain how spiral galaxies avoid the wind-

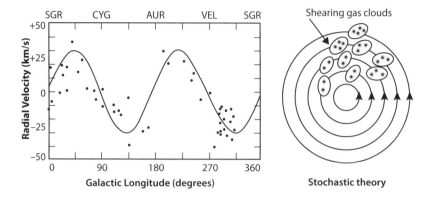

FIGURE 9.8. (*Top left*) Evidence for differential rotation (shearing) in the Milky Way comes from the run of stellar and nebular radial velocities with respect to galactic longitude. If the disk was rigidly rotating, everything in the disk would be fixed with respect to everything else, and so the measured radial velocities would all be near zero. Instead, they follow a double sinusoidal pattern of positive and negative velocities that is consistent with the differential motion. Near the Sun, the shearing amounts to about 8 kilometers per second per 1,000 light-years of galactocentric distance. (*Top right*) Schematic of randomly sited star-forming activity that gets sheared by the galaxy's differential rotation into spiral arm fragments. (*Bottom*) Examples of galaxies that may be hosting this sort of "stochastic" star-forming activity and spiral arm generation include the "flocculent" spirals NGC 4414 (left) and NGC 5055 (M63) (right). [(*Top left*) Adapted from *Introductory Astronomy & Astrophysics*, 4th edition by M. Zeilik and S.

ing dilemma is to abandon the assumption that the spiral arms are material constructs. Instead, the arms are tracing wavelike enhancements in density and star-forming activity. By substituting material arms with ephemeral pileups that propagate around the galaxy at a stately "pattern speed," astrophysicists have been able to explain many observed attributes of the so-called grand-design spiral galaxies.

Imagine a three-lane highway with a road crew in the slow lane. The crew is painting the line that marks the breakdown lane. Their painting machine allows them to make slow progress down the highway. To avoid hitting them, cars must vacate the slow lane and merge into the other two lanes. That process inevitably slows the traffic near the road crew—resulting in a bunching up of cars as their drivers endeavor to pass through this slowly moving bottleneck. This familiar scenario has many similarities to that of stars and gas passing through a slowly propagating density wave in a galactic disk. Although the analogy is not perfect, it helps us to understand how the spiral arms can manifest concentrated matter, but need not consist of the same material over time. Because the spiral arms delineate crests in the propagating density waves, they are no more permanent than the plodding bottleneck of cars. As long as the wave itself does not wind up, the spiral pattern can be maintained for billions of years.

Density wave dynamics provide compelling explanations for many of the observed properties found in "grand-design" spiral galaxies (see figure 9.9). These include resonant behavior at particular radii, where—in some spiral galaxies—ring-like pileups of gas and star-forming activity can be seen. Perhaps the pseudo-ring of molecular gas interior to the solar orbit is tracing such a reso-

A. Gregory, Orlando, FL: Saunders College Publishing–Harcourt Brace & Co. (1998), p. 387. (*Top right*) Adapted from *Galaxies and the Cosmic Frontier* by W. H. Waller and P. W. Hodge, Cambridge, MA: Harvard University Press (2003), p. 29. (*Bottom left*) The Hubble Heritage Team, Space Science Telescope Institute, Associated Universities for Research in Astronomy, NASA. (*Bottom right*) B. Hugo, L. Gaul, and A. Block, Advanced Observer's Program, National Optical Astronomy Observatories, Associated Universities for Research in Astronomy, National Science Foundation]

nance. Then there are the noncircular "streaming motions" that the gravitating arms induce on the orbiting stars and gas. Although we are hard-pressed to isolate these motions in the Milky Way, we can track them in other spiral galaxies such as M81 (see figure 9.6).

Perhaps most striking are the wholesale transformations of gas into newborn stars that occur in the spiral arms. As gas clouds approach the density-wave crests, their respective orbits crowd together, yielding enhanced rates of cloud-cloud collisions. The resulting buildup of larger and more pressurized clouds gives rise to greater rates of star formation. All this can be seen in nearby spiral galaxies, where dust lanes along the inner parts of spiral arms trace concentrations of molecular gas, crimson HII regions manifest the formation of ionizing O-type stars farther downstream, and blue starlight indicates the more evolved stellar populations that have lost their O-type stars but still include blue-white non-ionizing B- and A-type stars (see figure 9.9 and plate 15). Revisiting the analogy of the galactic garden, we would recognize in the "grand-design" spiral galaxies more formal arrangements reminiscent of the Jardin des Tuilleries and other French gardens.

Into the Halo

Given all this orchestrated star-forming activity along the putative spiral arms of our Galaxy, one might expect at least some of the associated pyrotechnics to vent into the halo. Astronomers have imagined possible scenarios, whereby multiple supernovae explosions from clusters of newborn stars blast through the disk—exhausting salvos of freshly forged heavy elements into the rarified "atmosphere" of our Galaxy (see figure 9.10). Evidence for such effervescent activity was at first hard to come by, but with the advancement of multi-wavelength astronomy, we now have compelling reasons to think that the spiral arms of the Milky Way are violently outgassing like the Pacific basin's volcanic "ring of fire."

Our first hints of the Milky Way's heavy breathing come from views at radio and infrared wavelengths, where a froth of nebular

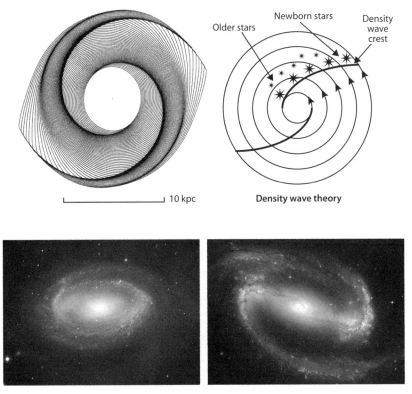

FIGURE 9.9. (*Top left*) Model of orbital streamlines in the presence of a two-armed spiral density-wave pattern. Near the density-wave crests, the stars and gas experience changes in trajectory that result in orbital crowding. (*Top right*) Schematic for gas clouds entering the crest of a spiral density wave whose gravitating dynamics induce the clouds to aggregate and spawn clusters of newborn stars downstream. These clusters continue to evolve and dim as their most massive stars quickly expire. (*Bottom*) Examples of "grand-design" spiral galaxies that are likely hosting spiral density waves include the resonant-ringed galaxy NGC 4725 (*left*) and the barred galaxy NGC 1300 (*right*).[(Top left) Adapted from H.C.D. Visser, "The Dynamics of the Spiral Galaxy M81–Part Two–Gas Dynamics and Neutral Hydrogen Observations," *Astronomy & Astrophysics* 88 (1980) 159. (*Top right*) Adapted from *Galaxies and the Cosmic Frontier* by W. H. Waller and P. W. Hodge, Cambridge, MA: Harvard University Press (2003), p. 29. (*Bottom left*) Rick Needham and Flynn Haase, Advanced Observers Program, National Optical Astronomy Observatories, Associated Universities for Research in Astronomy, National Science Foundation. (*Bottom right*) Hubble Heritage Team, Space Telescope Science Institute, Associated Universities for Research in Astronomy, NASA, and P. Knezek (Wisconsin-Indiana-Yale-NOAO [WIYN] Consortium, Inc.)]

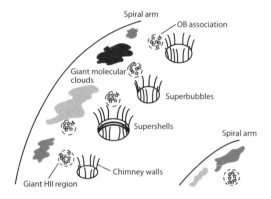

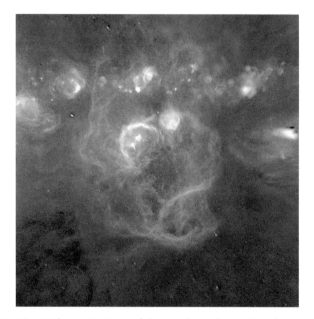

FIGURE 9.10. (*Top*) Schematic views of clustered star formation along spiral arms in the Milky Way venting into the galactic halo. Combined supernova blasts from expiring massive stars create "chimneys"—nebular pileups consisting of molecular, neutral, and ionized gas along with plenty of dust. Inside these chimneys, hot X-ray emitting gas is launched into the halo. (*Bottom*) The Orion-Eridanus superbubble spans an incredible 50 degrees in the sky. This view of ionized hydrogen emission from the superbubble features the brilliant (overexposed) Orion nebula and Horsehead nebula left of center, Barnard's Loop that surrounds these two H II regions, a smaller bubble inflated by the hot binary star system Lambda Orionis to the right, and other star-forming regions along the Milky Way above these objects.

filaments and shell fragments delineate major bulk motions (see plate 6 and figure 3.7). The regions surrounding the Cygnus, Carina, Orion, and Vela star-forming sites appear to be especially active. One of the closest of these blowouts is the "Orion-Eridanus superbubble." Spanning 50 degrees of the sky at a distance of 1,200 light-years, this eruption has displaced 1,000 light-years of galactic real estate. The excavation likely involved six or seven supernova explosions over the past 5–10 million years (see figure 9.10). Inside the superbubble's warm walls of atomic and ionized gas, X-ray emitting gas at temperatures of 1–10 million Kelvins is driving the bubble to ever greater dimensions. Indeed, there seems to be a hole in the bubble through which hot gas is leaking into the galactic halo.

Another notable extraplanar feature is the "Ophiuchus superbubble" which has been detected in the 21-centimeter line of atomic hydrogen at a distance of 25,000 light-years. Extending more than 25 degrees from the galactic midplane, the Ophiuchus superbubble is thought to span more than 11,000 light-years, to contain several million solar masses of neutral and ionized gas, and to require the equivalent of 100 supernova explosions.

At ultraviolet wavelengths, spectroscopic observations of remote but UV-bright quasars have revealed absorption features from "local" carbon atoms that have lost 3 of their electrons (CIV) and from oxygen atoms that are missing 5 of their electrons (OVI). These spectroscopic signatures of 1–10 million-degree gas are thought to trace a "coronal" component of extremely hot and rarified gas coursing throughout the galactic halo. Recent far-ultraviolet

---

[(*Top*) Adapted from C. A. Norman and S. Ikeuchi, "The Disk-Halo Interaction: Superbubbles and the Structure of the Interstellar Medium," *Astrophysical Journal* 345 (1989): 372–383. (*Bottom*) Made from the VTSS/SHASSA/WHAM H-alpha surveys of the sky, with reference to D. P. Finkbeiner, "A Full-Sky H-alpha Template for Microwave Foreground Prediction," *Astrophysical Journal Supplement* 146 (2003): 407. Data accessed by W. H. Waller from the SkyView interactive archive at NASA's Goddard Space Flight Center (see http://skyview.gsfc.nasa .gov)]

images at high galactic latitudes have revealed *emission* from these same highly ionized species (see figure 3.9).

If we could view the Milky Way from afar, what would all this eruptive commotion look like? Some clues come from multi-wavelength observations of nearby edge-on spiral galaxies (see figure 9.11). At optical wavelengths, vertical tendrils of dust and H-alpha emitting gas mark the "chimney walls" of the starburst plumes. X-ray images show more extensive halos of hot gas enveloping the stellar disks. These edge-on prototypes justify the notion that the Milky Way likely hosts a torrid seething atmosphere of its own making.

From our Earth-bound perspective, "what goes up must come down." If this old adage holds true on the galactic stage, we should find gas raining back down on the disk of the Milky Way. This so-called *galactic fountain* of violently expelled hot gas followed by cooler infalling gas has been predicted since the 1980s. However, the cooler component of the fountain has yet to be definitively confirmed. Instead, astronomers have found enormous clouds of hydrogen gas streaming through the halo. These clouds do not partake in the common rotation of the disk, and so relative to the Sun, stars, and clouds in the disk, they appear to be traveling at high speed—often on incoming trajectories. First discovered by radio astronomers in the 1960s, the high-latitude clouds were dubbed High Velocity Clouds (or HVCs) because of the odd Doppler-shifting of their hydrogen spectral-line emission. Subsequent investigations of their absorptive effects on the light from distant stars and from even more distant quasars has placed most of them in the outskirts of the Milky Way's extended halo (see figure 9.12). Spectroscopy of the background sources has shown that the absorbing gas lacks the heavy elements that characterize the clouds in the disk. This paucity of metals suggests they were not burped up by the disk but rather were formed as part of the galactic halo more than 10 billion years ago, or were stripped off hapless metal-poor companion galaxies more recently. A prime example of the latter scenario is the Magellanic Stream. Expelled by eruptive starburst activity in one or both of the Magellanic Clouds, the gas of the Magellanic stream has been tidally distorted by the

FIGURE 9.11. Some edge-on spiral galaxies exhibit extraplanar gas and dust that have been expelled from the galaxian disks. (*Top*) NGC 891 at optical wavelengths sports vertical tendrils of obscuring dust. (*Bottom*) NGC 3079 recently hosted a major outburst from near its nucleus. [(*Top*) C. Howk (Johns Hopkins University), B. Savage (University of Wisconsin), N. A. Sharp (National Optical Astronomy Observatories), Wisconsin-Indiana-Yale-NOAO (WIYN) Consortium, Inc., Associated Universities for Research in Astronomy, National Science Foundation. (*Bottom*) G. Cecil, University of North Carolina, Space Telescope Science Institute, Chandra X-ray Center, NASA]

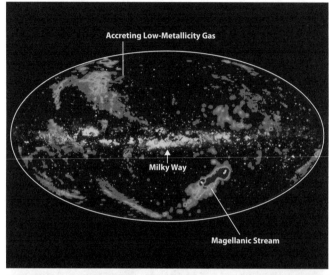

FIGURE 9.12. (*Top*) High-velocity clouds in the halo of the Milky Way as mapped in the 21-cm spectral line of atomic hydrogen. The Magellanic Stream extends from the Large and Small Magellanic Clouds (located to the lower right of this all-sky rendering) over a full quadrant of the southern sky. (*Bottom*) Smith's Cloud spans about 15 degrees of the northern sky. Compared to the other high-velocity clouds, it is much closer to the disk of the Galaxy and is expected to collide with the disk in about 20–40 million years. [(*Top*) Adapted from image composite by Ingrid Kallick of Possible Designs, Madison, Wisconsin, which is archived at http://hubblesite.org/newscenter/archive/releases/1999/1999/46/image/a/. The background Milky Way image is a drawing made at Lund Observatory. High-velocity clouds are from the survey done at Dwingeloo Observatory (A.N.M. Hulsbosch and B. P. Wakker 1988, "A Deep, Nearly Complete,

Milky Way's gravity over the past billion or so years. Today it sweeps across more than 90 degrees of the southern sky.

Closer to home, Smith's Cloud appears to be heading for an imminent collision with the disk of the Milky Way. Astronomers place this cool hydrogen cloud 40,000 light-years away in the direction of Aquila. At this distance, the cloud measures 11,000 light-years long and 2,500 light-years wide in a cometary configuration that suggests it is already interacting with diffuse gas in the galactic halo. It likely contains more than a million Suns worth of atomic hydrogen gas. Only 8,000 light-years from the galactic disk, Smith's Cloud is on a high-speed trajectory that should cause it to plow through the galactic disk at a 45-degree angle in the next 20–40 million years. Clouds like these give special credence to the idea that the Milky Way was rapidly built up through the accretion of ambient parcels—and is continuing to grow at a much slower rate today.

Meanwhile, astronomers keep looking for the elusive "rain" from the putative *galactic fountain.* The smaller population of intermediate-velocity clouds (IVCs) that have been discovered may end up filling the bill, as the IVCs appear to dwell closer to the galactic disk and to contain greater abundances of heavy elements. That would suggest they originated in the disk, where metal abundances are more like that of the Sun. Ejected into the halo by some coordinated collection of supernova blasts, they are now infalling as much cooler blobs of neutral atomic gas.

Cosmic Connections

All this galactic gardening bespeaks remarkably close relations among the diverse structural components that comprise the Milky

---

Survey of Northern High-velocity Clouds," *Astronomy & Astrophysics Supplements* 75 (1988): 119. (*Bottom*) Illustration by B. Saxton (National Radio Astronomy Observatory, Associated Universities Incorporated, National Science Foundation) based on research by J. Lockman and G. I. Langston (NRAO, AUI, NSF)and R. A. Benjamin and A. J. Heroux (University of Wisconsin–Whitewater)]

Way. As the pioneering American naturalist John Muir wrote in 1919, "When we try to pick out anything by itself, we find it hitched to everything else in the universe." Today, this sense of cosmic interconnectivity dominates our view of life within and among the galaxies—including our own Milky Way.

To get a better "feel" for your own personal stake in the Galaxy, just take a deep breath and think about it. Contained within that single breath are some 10 billion trillion (10^{22}) atoms of nitrogen, carbon, and oxygen that were first cooked up in the bellies of stars more than 4.6 billion years ago. In their expressive deaths, these stars spiced the Galaxy with heavy elements like so many drops of Tabasco in a roiling pot of gumbo. Today, you and I are living, breathing proof of the Milky Way's vigorous fecundity.

In the next chapter, we will turn our gaze to the galactic center—where the interactive dynamics that link ourselves to the rest of the Milky Way become ever more extreme and consequential.

CHAPTER 10

· · ● · ·

MONSTER IN THE CORE

The offing was barred by a black bank of clouds, and the tranquil water-way
leading to the uttermost ends of the earth flowed somber under an overcast
sky—seemed to lead into the heart of an immense darkness.

He cried in a whisper at some image, at some vision—he cried out twice,
a cry that was no more than a breath—"The horror! The horror!"
—From *Heart of Darkness* by Joseph Conrad (1902 CE)

LIKE SOME SLEEPING GIANT, the core of our Galaxy softly snores beneath thick blankets of dust. This somnolent situation was not always the case, however, nor will it be in the future. Just consider what lurks within the central light-year, and you can begin to appreciate the incredible power inherent to the Galaxy's dark heart. In this chapter, we will survey (from a safe distance) the many unique phenomena that have been found in the general direction of the galactic center. We will then vicariously zoom into the core as far as our observations permit, assess the case for a supermassive black hole therein, and last, speculate upon the possible environmental impacts of this central "monster" upon the rest of the Galaxy.

The Galactic Bulge/Bar

A June stroll on a clear, moonless, un-illuminated night anywhere south of the equator would present to you the Sagittarius, Scorpius, and Ophiuchus star clouds in all their glory. Bulging above and below the plane of the dust-riven Milky Way, the diffuse star-light would immediately impress even the most casual of observers to conclude that this part of the sky contains something very special. The galactic center itself is hopelessly obscured behind multiple clouds of gas and dust within the intervening disk of the Milky Way. Fewer than one optical photon in a trillion makes it out from the center as far as the solar orbit. However, a few degrees above and below the murky disk, the dust begins to thin out—permitting relatively clear views of the stars that comprise the bulge.

Especially clear "windows" have been identified, the most famous of which is Baade's window—named after the twentieth-century German and American astronomer Walter Baade. Located at 1.04 deg galactic longitude and –3.88 deg galactic latitude, Baade's window, along with the half-dozen other windows that have been identified, have served as precious portals for astronomers keen on characterizing the stars of the inner bulge (see figure 10.1). Beyond these exceptional sight lines, the more conspicuous outer bulge extends approximately 20 degrees in longitude and 10 degrees in latitude away from the galactic center. If one adopts a distance to the galactic center of 27,700 light-years (8.5 kpc), the bulge region spans some 20,000 light-years in diameter and 10,000 light-years in thickness. This amounts to roughly one-fifth of the galactic disk's total diameter, but more than four times the disk's thickness.

As noted before, mapping at infrared wavelengths has revealed asymmetries in the overall shape of the inner bulge, suggesting to astronomers that the bulge is itself oblong—a knockwurst-like "prolate spheroid" that is three times longer than it is wide. Many astronomers prefer to think of a bar that is inclined to our line of sight by 30–45 degrees. Others allow for a bulge/bar hybrid. This distinction is not merely semantic, however, as bulges and bars are thought to derive from completely different evolutionary processes. The classical bulges observed in other spiral galaxies are commonly attributed to the ancient spheroidal or halo component, where each bulge manifests the densest concentration of halo stars. By contrast, galactic bars are thought to belong to the disk compo-

FIGURE 10.1. Baade's window is an exceptionally unobscured sightline toward the central bulge/bar of our galaxy. Inside it, the ancient globular clusters NGC 6522 (upper) and NGC 6528 (lower) highlight a rich stellar tableau. [Adam Block, Mount Lemmon SkyCenter, University of Arizona]

nent, where resonances between orbiting density waves and stars in the disk have led to the prolate pileups. This ambiguity of origin for the galactic bulge/bar remains unresolved.

Spectroscopic studies of the stellar populations in the bulge/bar have yielded fascinating insights on the composition, kinematics, and ages of these stars. Most of the stellar spectra are of evolved giant stars, as the more abundant main-sequence stars are too faint to be studied in such detail. These spectra indicate red giant and asymptotic giant branch stars of near-solar masses and roughly 10–12-billion-year ages. However, there is also amazing diversity in the metal abundances and kinematics of the bulge/bar stars. The metallicities range from ten times solar to less than 1/100 solar, while the rotation rate of the bulge/bar stars appears intermediate between those of the sprightly disk and the sluggish halo. Even the ages are far from fixed, with compelling spectroscopic evidence for a peppering of much younger stars having taken up residence in this congested region. Our overall picture is of an ancient metropolis that has acquired many layers of history and is continuing to host new life in an ongoing reinvention of itself. Rome comes to mind.

Perhaps the greatest argument for this ancient picture is that the bulge/bar of our Galaxy hosts a supermassive black hole in its center. In observations of other large galaxies, astronomers have found a remarkably tight correlation between the masses of the central black holes and the total stellar masses in the surrounding spheroids (bulges and halos). The correlation yields black hole masses that are roughly 0.2 percent (1/500) of the corresponding spheroid masses. This remarkable relation suggests to astronomers that both the black holes and the spheroids were likely built together as part of the same overall process. The preferred epoch for making these sorts of constructs can be handily traced by the powerful quasars, whose freshly forged black holes were then ravishing their immediate surroundings and blazing forth at all wavelengths. We see the greatest numbers of quasars at spectral redshifts of about 2–3 which correspond to lookback times of 9–10 billion years.[1] Because our Galaxy hosts one of these supermassive black holes, the "monster" and its enfolding spheroid (including the bulge) were likely built together some 10–12 billion years ago. How the emergent black hole affected the growth of the spheroid—and vice versa—remains hotly debated. We will revisit this debate in the next chapter, when we consider scenarios of the Milky Way's birth and subsequent evolution over cosmic time.

The Inner 1,000 Light-years

Once we restrict our regard to the central 2×2 degrees ($1,000 \times 1,000$ light-years) that surround zero galactic longitude and latitude, we are entering the realm of the busy and bizarre. Mapping of radio continuum emission from this region has revealed a hodge-podge of emission nebulae powered by clusters of newborn stars, supernova remnants that are expanding into unusually pressurized surroundings, and strange "threads" of excited gas that appear to leap from the disk (see figure 10.2). The view at X-rays is equally bizarre, with thousands of hot point sources and patches of diffuse high-energy emission that beg physical explanation (see plate 16).

The accumulation of stars nestled within this nebular morass is nothing less than phenomenal. Near-infrared imagery (see figure 9.5) and follow-up spectroscopy to determine spectral types and distances yield total dimensions of about 300 light-years and a mass exceeding 100 million Suns—at least 100 times the heft of any other cluster in the Milky Way. We would be remiss in calling this colossal stellar pileup a bona fide cluster, however, as it was likely cobbled together over several billions of years rather than the usual million or so years. Indeed, this most nucleated part of the bulge/bar contains stars that probably formed elsewhere in the disk and have since fallen into the galactic center through a variety of tidal provocations. Theoretical astronomers have surmised that stellar bars inevitably drive gas and stars toward the centers of their host galaxies—and that the growing central protuberances ultimately destroy the bar "instabilities." In the Milky Way, we may be witnessing an intermediate stage in this ongoing saga of bar-induced central consolidation, bar destruction, and regeneration.

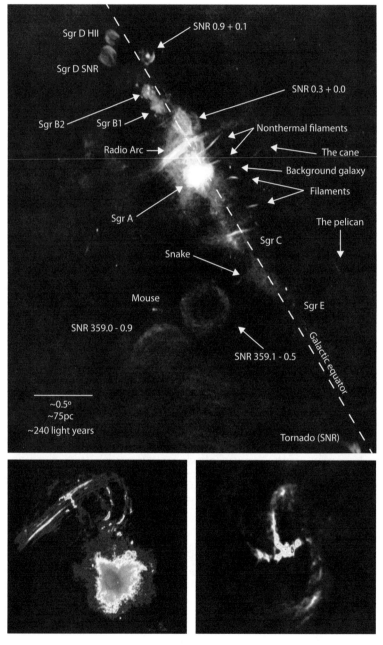

FIGURE 10.2. (*Top*) Radio continuum emission from the central 3° x 4° of the Galaxy. At an assumed distance of 27,700 light-years, this corresponds to roughly 1,500 x 2,000 light-years. The radio emission is rendered in equatorial

The Central Hotbed—Sagittarius A

As we zoom in closer to the apparent center of galactic activity, our interpretive capabilities become increasingly ungrounded. Nothing else in the Milky Way has prepared us for what we can now perceive. Through the careful combination of signals from multiple radio antennas (in a process known as interferometry), astronomers have been able to map the central 100 light-years at exquisitely high angular resolution. The resulting vista of radio-continuum emission reveals a bright bulbous splotch of activity (known as Sagittarius A) that appears to connect with a crazy streak of striated nebulosity (see figure 10.2). The long, straight upper part of the so-called arc traces synchrotron emission from relativistic electrons that have been compelled by intense magnetic fields to twirl and radiate away their energy. The source of all this energy remains uncertain—it could be as "mundane" as an influx of electrons from the reconnection of magnetic fields associated with jostling molecular clouds in the area. Or perhaps one or more supernova blasts

---

coordinates, with the disk of the Milky Way running from lower right to upper left. (*Bottom left*) Sagittarius A and filamentary arc in the galactic center as mapped in the radio continuum at a wavelength of 20 cm. The field of view is 1/10 that shown above (0.42° x 0.42° or 200 x 200 light-years). The long, straight streak of filamentary emission is directed perpendicular to the galactic plane. Inside Sgr A, the dark oval toward the left of the galactic center traces the supernova remnant Sgr A East. (*Bottom right*) Close-up of Sagittarius A as mapped in the radio continuum at a wavelength of 3.6 cm. The field of view is another 15 times smaller (1.7 x 1.7 arcminutes or 13.7 x 13.7 light-years). This field features the "mini-spiral" of ionized gas known as Sgr A West. The bright central source in Sgr A West marks the still unresolved nuclear source Sgr A*, where a supermassive black hole is thought to reside. [(*Top*) N. E. Kassim, D. S. Briggs (NRL), J. Imamura (TJHSST), T.J.W. Lazio (NRL), T. N. LaRosa (Kennesaw State University), and S. D. Hyman (Sweet Briar College), Remote Sensing Laboratory, Naval Research Laboratory. (*Bottom left*) F. Yusef-Zadeh (Northwestern University), M. R. Morris (University of California, Los Angeles), and D. R. Chance, National Radio Astronomy Observatory (NRAO), Associated Universities Incorporated (AUI), and the National Science Foundation (NSF). (*Bottom right*) F. Yusef-Zadeh (Northwestern University), D. A. Roberts and W. M. Goss, National Radio Astronomy Observatory (NRAO), Associated Universities Incorporated (AUI), and the National Science Foundation (NSF)]

have helped to shape and power this remarkable feature. Then again, one could imagine a more exotic scenario involving a prior outburst from the supermassive black hole that is thought to lie deep within Sagittarius A. To date, the more mundane interpretation appears to best explain the available observations.

Unlike the synchrotron-emitting upper arc, the nebular filaments (or "Arches") that appear to link the radio arc to Sagittarius A emit a thermal spectrum characteristic of hot ionized gas. Here, astronomers are more confident of the powering mechanisms, as several robust clusters of newborn stars have been found in this area. The most prominent of these are the Arches and Quintuplet clusters—both of which are extraordinarily rich in the most massive of stars. The scorching ultraviolet emission and mighty winds from the hundreds of O-type and Wolf-Rayet stars that are churning therein would be fully capable of ionizing and sculpting the observed filaments.

Sagittarius A itself has been known since 1951, when it was first identified as the brightest radio source in the constellation of Sagittarius. Subsequent mappings at higher resolution have shown it to consist of two key components. Sagittarius A East, the larger and fainter of the two, has all the trappings of being a supernova-driven shell. The non-thermal spectrum of its radio emission is consistent with a synchrotron radiation mechanism involving intense magnetic fields and relativistic electrons that were likely accelerated by a supernova blast about 10,000 years before the observed epoch.

Compared to other radio-emitting supernova remnants of similar type in the Galaxy, Sgr A East is unusually small—only about 30 × 20 light-years as seen in projection. Astronomers have inferred from this size disparity that Sgr A East is plowing into an especially thick and resistive interstellar medium. Dense molecular clouds have been found in the immediate vicinity of this supernova remnant which together may help to explain its retarded expansion. Inside the radio-emitting shell, a haze of X-ray emission has been detected—the consequence of shock heating from blast waves that have bounced off the shell walls and rippled throughout the interior of Sgr A East.

More detailed X-ray observations of the central 8.5 × 8.5 arc-minutes (68 × 68 light-years) by the Chandra X-ray Observatory have yielded even stranger incarnations of the tumultuous activity in this densely packed region (see plate 16). On large scales, two giant plumes of 20-million-degree gas appear to be billowing outward from Ground Zero in opposite directions. Each plume measures about 30 light-years and contains about a solar mass of superheated gas. These dimensions and energetics suggest to astronomers that some expulsive event must have occurred thousands of years before the 27,700-year-old epoch that we are currently witnessing (where we must account for the light-travel time from the galactic center to the Solar System). A prime candidate for feeding such an eruption would be any star that happened to stray too close to the nuclear black hole. The intense tides immediately surrounding the black hole could have torn the star asunder and siphoned its remains onto the circumnuclear accretion disk, thus inducing an explosion of superheated and highly magnetized gas perpendicular to the disk.

On smaller scales of about a light-year, the X-ray emission is structured into strange arcs and filaments that together suggest a many-petaled flower or frilly sea creature (see plate 16). Astronomers have puzzled over these odd features, concluding that they delineate the compressed interfaces between colliding stellar and pulsar winds. Given the uncanny concentration of hot massive stars and interacting binaries in this area, such an interpretation appears well-founded. It also helps to explain Sgr A East as just one of many supernova remnants that have rocked the galactic center over the eons.

The Nuclear Maelstrom—Sagittarius A West

Superposed in front of Sgr A East swirls one of the most fantastic cosmic objects ever discovered. Sagittarius A West spans only about 14 light-years of galactic territory, but it sits squarely upon the true hub of the Milky Way. This radio source looks like a three-armed pinwheel or mini-spiral (see figure 10.2). However, our view of it may be a spurious juxtaposition, as we don't yet know whether the ensemble's components all share the same plane in the sky.

The "arms" of the apparent mini-spiral emit a thermal radio spectrum and so they most likely harbor warm ionized gas. A nearly identical pattern of infrared line emission from this region confirms the thermal nature of Sgr A West while helping astronomers to track the various motions. Analysis of the Doppler-shifted emission reveals both rotating and infalling motions that have prompted astronomers to wonder where the gas is coming from and where it is heading. The necessary gaseous reserves appear to have been identified, as millimeter-wave observations have delineated a blotchy ring of dense molecular gas that immediately surrounds the mini-spiral of ionized gas. The ultimate destination of this nebular flux is less clear, but at least some of it must occasionally find its way into the dynamical center of the Galaxy—and the supermassive black hole that most astronomers now think lies within.

The Black Hole's Lair—Sagittarius A*

From Sgr A West, one must zoom in by a dizzying factor of a thousand to eke out any information about the radio source that marks our Galaxy's ultimate nucleus—Sagittarius A*. This feat was first accomplished in 1974, when astronomers at the National Radio Astronomy Observatory in West Virginia pointed their newly commissioned array of four radio dishes toward Sgr A. The combined signals from these antennas revealed a discrete source that spanned less than 0.1 arcsecond of angle. If located in the galactic center some twenty-seven thousand light-years away, such a source would measure no more than 1/100 of a light-year (a few light-days) in diameter. Suddenly, we were dealing with something very unique that could not be explained by the usual invocations of intense nuclear starburst activity.

Subsequent mapping with increasingly powerful interferometric arrays of radio telescopes has further reduced the diameter of Sgr

A* to less than 1/10,000 of a light-year (the equivalent dimension of our Solar System's asteroid belt). The resolved size turns out to depend on the radio wavelength, with 20-cm observations yielding 1/60 light-year sizes and mm observations yielding the smallest limits. Astronomers have concluded from this behavior that we are viewing Sgr A* through a fog of its own making—and that this atmospheric scrim is what we are resolving at radio wavelengths.

Although Sgr A* remains an enigmatic radio source, observations at other wavelengths have provided vital clues to its physical nature and environmental circumstances. Immediately surrounding Sgr A* swarms an incredibly dense cluster of stars (see Figure 10.3). More than 400 stars occupy the innermost light-year of our Galaxy's nuclear cluster. Infrared spectroscopy of these stars shows that many of them are very hot, massive, and young—confirming the notion that the circumnuclear maelstrom of Sgr A West has been feeding star-forming gas into the Galaxy's nuclear core. One star—IRS 7, the brightest source in figure 10.3 (top)—appears to be experiencing windy provocations from the central source, as it exhibits a cometary extension at radio wavelengths.

The infrared observations have also allowed astronomers to track the motions of individual stars in the nuclear cluster. Over the course of several years, some of these stars have made complete orbits around the galactic center. Speeding at thousands of kilometers per second along highly elliptical trajectories, they appear to be orbiting the same focal spot—Sgr A*! One star—S0 16—careens within 90 astronomical units of Sgr A* (see figure 10.3, Bottom). From these exquisitely delineated stellar motions, astronomers have used the basic orbital elements to calculate the gravitating mass that has induced this dynamic scene. The resulting mass of 4 million Suns within a radius of no more than 90 astronomical units leaves very few options in the way of explanation. Sgr A* is either harboring a supermassive black hole or an unfathomably dense cluster of stars that will very soon become a supermassive black hole. Indeed, the black hole option turns out to be the more "stable" solution. An incredible conclusion perhaps, but so far unavoidable.

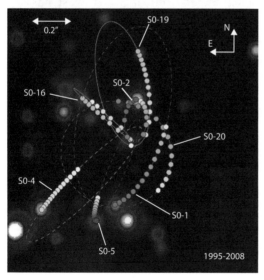

FIGURE 10.3. (*Top*) The nuclear cluster of stars that occupies the central 23 arcseconds (3 light-years) surrounding Sgr A*. Many of these stars are very hot, massive, and short-lived. Therefore, a recent burst of star formation must have occurred in this extraordinarily dense region. (*Bottom*) Orbital paths of stars within 1 arcsecond (0.13 light-year) of Sgr A*. These motions (as mapped between 1995 and 2008) have yielded a dynamical mass for Sgr A* of about 4 million Suns—all concentrated within a perimeter equivalent to the orbit of Pluto around the Sun. [(*Top*) S. Gilleson et al., Very Large Telescope, European Southern Observatory, with reference to S. Gilleson et al., "Monitoring Stellar Orbits around the Massive Black Hole in the Galactic Center," *Astrophysical Journal* 692 (2008): 1075–1109. (*Bottom*) A. M. Ghez et al., UCLA Galactic Center Group, University of California, Los Angeles with reference to A. M. Ghez et al., "Stellar Orbits around the Galactic Center Black Hole," *Astrophysical Journal,* 620 (2005): 744]

Enigmas at the Brink of No Return

Black holes are black because their gravitational grasp keeps any light from escaping their immediate vicinity. By equating the gravitational and kinetic energies associated with each photon of light, astronomers have derived a simple expression that predicts the effective size of a black hole. The so-called Schwarzschild radius of a nonrotating black hole depends solely on the black hole's mass according to

$$R_S = 2\ G\ M\ /\ c^2,$$

where R_S is the radius of the "event horizon," inside of which no light can escape, G is the constant of universal gravitation, M is the mass of the black hole, and c is the speed of light. For a black hole containing the equivalent of 4 million Suns, the corresponding Schwarzschild radius would be about 12 million kilometers, or about one twelfth of an astronomical unit (five times *smaller* than Mercury's orbital radius around the Sun!). This would subtend only 10 micro-arcseconds at the distance of the galactic center. Though minuscule, such an angular dimension may soon become within reach of our largest radio interferometers. In particular, an intercontinental Very Long Baseline Interferometer (VLBI) operating at millimeter wavelengths should be capable of resolving the event horizon of our Galaxy's nuclear black hole. Theorists have modeled what might be seen and have concluded that the black hole would act as a powerful gravitational lens—distorting any background sources of radio emission into an arc of light that would wrap around the event horizon and associated accretion disk.

Meanwhile, we cannot help but puzzle over the central engine's current docility. Compared to other galactic nuclei that have been shown to harbor a supermassive black hole, our galactic nucleus appears to be strangely quiescent. Radio luminosities are down by factors of a billion, and X-ray luminosities barely rise above that of a single supernova remnant. Perhaps we are catching our nucleus in a pronounced lull. To be sure, radio, infrared, and X-ray astronomers have found the Galactic nucleus to dramatically vary in its radiant output on timescales of months, days, and hours. However, the overall level of activity is decidedly dim.

We are left to wondering what the galactic nucleus was once like. Could it have gone through a "quasar" phase, where it blazed

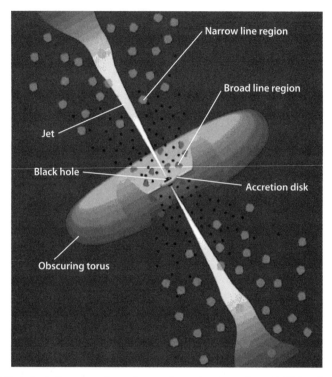

FIGURE 10.4. Artist's conception of an active galactic nucleus (AGN), where the supermassive black hole is surrounded by a brilliant accretion disk that is pow-ered by infalling matter. Some of this matter settles into a dense obscuring torus well beyond the accretion disk, while other matter is processed by the accretion disk into a highly collimated bipolar outflow of relativistic gas. The broad-line and narrow-line regions refer to blobs of gas that have been irradiated by the ultra-hot accretion disk to glow in characteristic spectral emission lines. The Milky Way's nucleus may have all of these components, but in a currently quies-cent state. [C. M. Urry (Yale University) and P. Padovani (Space Telescope Sci-ence Institute), High Energy Astrophysics Science Archive Research Center (HEASARC), Goddard Space Flight Center, NASA]

with the equivalent of a trillion Suns, far outshining the entirety of its galactic surroundings? When we examine other currently active galactic nuclei (AGNs), we see the hallmarks of a supermassive nucleus ravenously feeding on its immediate surroundings, radiat-ing well beyond its gravitational limits, and spewing its excess fuel in the form of vast bipolar jets (see figure 10.4). These sorts of pyrotechnics were certainly possible for our galactic nucleus, when it was still surrounded by lots of infalling matter. Given our Galaxy's pronounced central bar and its associated asymmetric gravitational field, there may come another day when a resurgence of central fueling will give rise to a new epoch of brilliant activity.

What then? We can only speculate upon the few clues that are available to us. For those rare galaxies currently hosting powerful nuclear activity, we can see that the tremendous luminosities and outflows reach well beyond the black holes' immediate environs. Some astrophysicists propose that such powerful activity can heat up the interstellar contents of the host galaxies, quenching any further star formation while promoting wholesale evaporation of the gaseous stores. This may help to explain the population of barren elliptical galaxies that are known to harbor supermassive black holes. Perhaps they were once like the Milky Way, but have since lost their gaseous contents through both internal heating by their central black holes and external disruption through encounters with other galaxies. Blazing and erupting galactic nuclei may also account for the early stifling of rampant star formation within the bulges of spiral galaxies like the Milky Way. Indeed, the feedback provided by intense nuclear activity may have induced the tight correlation that is observed between the bulge and black hole masses. To address these sorts of conjectures, we find ourselves pondering the full sweep of galactic evolution—the topic of the next chapter.

·· • ●● ··

TALES OF ORIGIN

So many atoms, clashing together in so many ways as they
are swept along through infinite time by their own weight, have
come together in every possible way and realized everything
that could be formed by their combinations.
—Titus Lucretius Caras in *On the Nature of Things* (55 BCE)

In order to make an apple pie from scratch,
you must first create the universe.
—Carl Sagan in *Cosmos* (1980 CE)

HOW FAR BACK IN TIME must we go in order to perceive the Milky
Way's origins? If our estimated ages for various constituents of the
Milky Way are correct, we pretty much have to go all the way back
to the Hot Big Bang itself, some 14 billion years ago (see figure
11.1). Consider the lowly hydrogen atom, the most abundant sub-
stance making up our Galaxy's stars and nebulae. Its single proton
was likely forged from a triplet of quarks around a millionth of a
second following the Big Bang—as the primeval fireball *cooled* to
a temperature of a trillion (10^{12}) degrees Kelvin. The proton's part-
nering electron emerged from the seething broth of particles and
antiparticles, when the universe had expanded for more than 1
second and had cooled to a relatively balmy billion (10^9) degrees.
To build the nucleus of the next most abundant element, helium,
the universe had to diffuse and deactivate enough for the necessary
fusion reactions to take hold. This occurred in a matter of minutes.
The centers of hydrogen-fusing stars provide a current-epoch anal-

ogy to the thermal conditions that prevailed during this important cosmic epoch. The characteristic densities then were much lower than those found in the thermonuclear cores of stars, however. The air above Mount Everest provides a closer match in density—just imagine this same air undergoing fusion reactions and you may get a sense of what it must have been like during the nucleosynthesis era.

Letting another 380,000 years pass, we see that the universe has decompressed and cooled to a state akin to that of an HII region such as the Orion and Rosette Nebulae. With temperatures less than 10,000 Kelvins and densities of a few hundred particles per cubic centimeter, the plasma of ions and free electrons could readily recombine to form neutral atoms. Suddenly, the photons were no longer subject to the incessant scattering off of the negatively charged electrons and positively charged ions in the plasma. Streaming past the neutralized atoms, the photons were free for the first time to pervade the expanding cosmos. The neutralizing process also allowed the ordinary matter to gravitationally congeal—unimpeded by the pesky photons. We see the liberated photons today as the all-encompassing Cosmic Microwave Background (CMB)—the so-called echo of the Big Bang (see figure 11.1). Though originally emitted in the form of visible light waves (which may have appeared like an omnipresent orange glow), this radiation has since been stretched by the expanding universe into the microwave portion of the electromagnetic spectrum. Today, space-borne radiometers can trace the primeval CMB with great precision throughout the sky.

Faintly imprinted upon the CMB are telltale irregularities with characteristic amplitudes of less than one part in 10,000 and with angular spacings of about a degree. Observational cosmologists have interpreted these barely legible runes in the CMB as fossil footprints of incipient structuring in the emergent cosmos. According to current reasoning, the irregularities trace relative overdensities and voids in the overwhelming dark matter, toward which the ordinary "baryonic" matter has flowed like fish to bait. We now have nodes of concentrated matter, within which the business of galaxy formation could proceed.

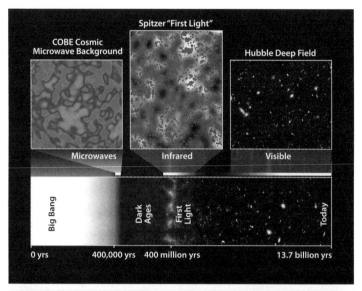

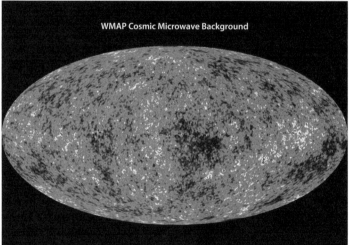

FIGURE 11.1. (*Top*) Schematic timeline of the universe with observations of key epochs featured. The Cosmic Microwave Background marks the epoch when the universe cooled to a neutral atomic state that was transparent to its own radiation. The infrared background mapped by the Spitzer Space Telescope is thought to be tracing the first starlight. The visible-light Hubble Deep Field contains galaxies with lookback times going back to the first few billion years after the Big Bang. (*Bottom*) All-sky rendering of the Cosmic Microwave Background (CMB) as mapped by the Wilkinson Anisotropy Probe (WMAP) after removal of foreground emission from gas in the Milky Way and adjustment for our galaxy's motion with respect to the cosmic reference frame. This radiation corre-

A Tale of Two Scenarios

Over the past four decades, two paradigms of galaxy formation have vied against one another in a seemingly unending competition for primacy. The "top-down" model came first, with its notion of galaxies emerging wholesale from the primeval irregularities that were then only predicted to exist. The "bottom-up" model came later, as evidence for the hierarchical assembly of dwarf galaxies into larger systems began to accrue (see figure 11.2). The "bottom-up" model also benefited from the notion of cold dark matter pervading the universe. Unaffected by the blaze of photons, the dark matter could lump up even before the epoch of recombination and so facilitate the structuring of galaxies early in the universe's history.

Consider the plasma state of the baryonic universe shortly before the grand recombination of electrons onto ions and subsequent neutralizing of all ordinary matter. Given the densities and temperatures that prevailed during this time, the most favored gravitational constructs had dimensions of a few tens of light-years and masses of a few million Suns. Anything smaller and less ponderous would have been unable to gravitationally congeal. Any significantly larger entities could not be ruled out, but were far less likely. So, we are left with flurries of proto-dwarf galaxies to build up the grand spiral and elliptical galaxies that we know and love today.

Within these nebular motes, the first generation of stars played itself out. Bereft of heavy elements that would otherwise have radiatively cooled the gas to tens of Kelvins, the first stars had to

---

sponds to an all-encompassing "surface of last scattering" with a temperature of 2.725 Kelvins and lookback time of 13.7 billion years. The bright patches trace relatively warmer and presumably less concentrated ordinary matter. The dark patches trace the cooler and denser domains. These barely perceptible irregularities in the microwave emission (of order 0.0002 Kelvin) are thought to trace large-scale structuring in the early universe. [(*Top*) A. Kashlinsky (NASA's Goddard Space Flight Center), Caltech, Jet Propulsion Laboratory, NASA. (*Bottom*) Wilkinson Anisotropy Probe (WMAP) Science Team, NASA]

form from much hotter gas with temperatures in the 200–300 K range. These temperatures were fixed via the relatively inefficient cooling mechanism of excited molecular hydrogen radiating at infrared wavelengths. In order for gravity to prevail over the thermal tumult at these temperatures, the first stars had to be very massive. The surfaces of the massive stars were very hot and so radiated prodigious amounts of extreme ultraviolet light into space— enough to completely re-ionize the universe! Observational support for a cosmic epoch of re-ionization having occurred between 400 million years and a billion years after the Big Bang continues to accrue. Telltale irregularities in the Cosmic Microwave Background have been attributed to excess scattering of the CMB radiation by free electrons following re-ionization. Also, the spectra of the most remote quasars show changes depending on whether the quasars are observed at lookback times before or after the re-ionization epoch.

The universe on intergalactic scales continues to be in a state of ionization. However, within the proto-galactic gas blobs, rapid changes were afoot. Once the first stars formed and had their days of blazing glory, they died in supernova explosions that seeded their host blobs with heavy elements. Radiative cooling by these various atomic species subsequently ramped up, thus allowing the gas to settle down into the more familiar states of cold molecular and cool atomic clouds that characterize today's dwarf galaxies.

As detailed observations of nearby dwarf galaxies have shown, the pools of hydrogen gas in these galaxies are swirling around much faster than can be explained by the gravitational bonding that is expected from the visible matter. Instead, astronomers have invoked large halos of cold dark matter that are responsible for keeping the itinerant gases from flying away. As the proto-dwarf galaxies merged into much larger galaxian realms, their dark-matter halos also merged. What we see today are the legacies of these dark and luminous mergers—self-organizing into the exquisite galaxian systems that define the local universe (see figure 11.2).

The above story continues to resonate among astronomers intent on understanding the emergence of discrete structures from the chaos of the Hot Big Bang. However, it also has ongoing issues.

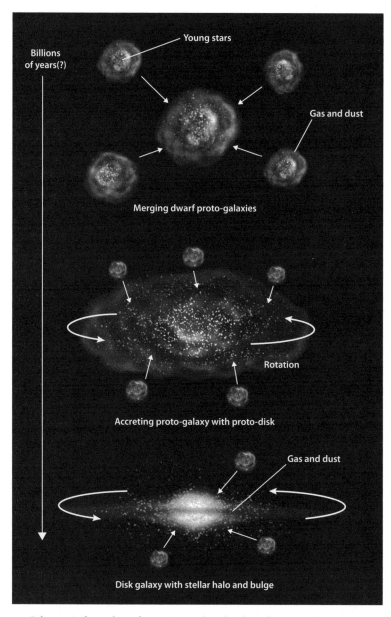

FIGURE 11.2. Schematic for galaxy formation, whereby dwarf proto-galaxies merge to create a giant spiral galaxy like the Milky Way. (*Top*) Merging proto-galaxies would likely have tidal tails and be traveling on in-spiraling trajectories. For simplicity, these aspects are not shown. (*Middle*) In this hierarchical formation scenario, the angular momentum of the combined systems is significant, leading to a rotating disk of gas that in turn spawns subsequent generations of stars. (*Bottom*) Young disk galaxy like the early Milky Way. Stars that formed in the proto-galaxies before they merged make up the swarming halo and much of the central bulge. [W. H. Waller and L. Slingluff based on multiple sources]

Consider what we have found from our most sensitive and detailed images of deep space. Arrayed throughout the famous Hubble Deep Field and even more sensitive Hubble Ultra-Deep Field, we can perceive myriad faint and apparently small galaxies (see figure 11.3). These findings would appear to support the idea that lots of dwarf galaxies merged to form the giant spiral and elliptical galaxies that we see today in the local universe. However, the light from these remote sources has been redshifted by the intervening expansion of the universe—transforming once ultraviolet emission into the visible blotches that were imaged by the Hubble Space Telescope. We are therefore privy to the light from these galaxies' hottest, most UV-bright star-bursting populations, yet are essentially ignorant of the redder, longer-lived stars that may trace out much larger galaxian systems. Some deep images at infrared wavelengths that keep up with the redshifting have in fact found significantly larger disk-like galaxies, thus calling into question the purely hierarchical scenario of galaxy construction via merging dwarfs. Instead, it appears that many disk-like galaxies—and perhaps even the inner parts of the Milky Way itself—were already recognizable within a few billion years following the Big Bang.

Another issue vexing astronomers is the prediction that giant spiral galaxies like the Milky Way should be surrounded by thousands of dwarf galaxies. These necessary vestiges from the cold-dark-matter-dominated scenario of hierarchical merging should now be detectable in deep all-sky surveys. Instead, no more than 40 dwarf galaxies are known to be associated with the Milky Way Galaxy. Even if the 150 known globular clusters in the galactic halo are included in the tally as former dwarf galaxies that have been consumed and stripped to their cores, we are still far from the predicted numbers. Perhaps we have yet to image deeply enough. Or perhaps the current bottom-up model needs serious revision. Fortunately, we can look forward to the next generation of ground-based and space-borne telescopes to delineate with much improved sensitivity the distribution of galaxian matter in both the remote and local universe. Some of these missions will be targeting the Milky Way itself, as our home galaxy has left behind a rich legacy of its former self.

FIGURE 11.3. The Hubble Ultra-Deep Field (HUDF) and an even larger HST survey of deep space have together revealed more than 500 galaxies that were formed when the universe was less than one billion years old. In this segment of the HUDF, four of the young star-bursting galaxies are marked, with close-ups shown below. These galaxies appear much smaller than the Milky Way and the other giant galaxies that characterize our current epoch. [Adapted from R. Bouwens and G. Illingworth (University of California, Santa Cruz), NASA and the European Space Agency. See http://hubblesite.org/newscenter/archive/releases/cosmology/2006/12/image/a/]

Galactic Archaelogy (Redux)

Over the eons, the Milky Way Galaxy has been stirred but not significantly shaken. Consequently, one can use the current distribution of stellar ages, motions, and chemical abundances to ascertain a fair bit of our Galaxy's natural history. We did some of this in chapter 5, when we first considered the stellar distributions in the disk and halo. But there is plenty more to mine, beginning with the glaringly obvious fact that the Milky Way sports a thin disk of stars, gas, and dust. Compared to the disk's 100,000 light-year diameter, the vertical distribution of old stars in the thin disk extends only 2,000 light-years from the midplane, while the thickness of young stars and molecular gas clouds is a paper-thin 300 light-years. Most of the ordinary matter in the Milky Way resides within this disk—perambulating around the center on mostly circular trajectories that share nearly the same orbital plane. This rather obvious defining feature of spiral galaxies need not have occurred. If the Milky Way had suffered a close encounter or merger with a galaxy of equivalent girth, the mutual scrambling of stellar orbits would have produced a giant elliptical galaxy. Indeed, this fate is predicted for the Milky Way about 4 billion years from now, when it and the Andromeda Galaxy get too close for comfort.

Left essentially to itself, the Milky Way has nonetheless suffered some galactic provocations that have made their marks. For example, the warping of the thin disk was likely induced via gravitational interactions with one or more smaller galaxies. Similarly, the so-called thick disk may embody the remainders of a relatively low-mass galaxy that mixed it up with the Milky Way approximately 10 billion years ago. Extending more than 3,000 light-years above and below the thin disk, but containing only 10 percent of the thin disk's stellar mass, the thick disk may in fact represent the residual stars that were stripped off a galaxy whose core is what we might now recognize as the giant globular cluster Omega Centauri (see figure 3.10). Alternatively, the thick disk may represent a residual stellar population remaining from the original epoch of disk formation or a fluffing up of stars and gas that was provoked entirely by internal dynamics within the thin disk (see chapter 5). Astronomers have found that the stars of the thick disk are typically older and more metal-poor than those in the thin disk (albeit with wide variations). This suggests an early formation scenario that likely preceded the thin disk. The infall of one or more preexisting dwarf galaxies could both explain the thick disk's hoary age and account for its typically lower metallicities.

Beyond the thick disk, the galactic halo hosts a swarm of globular clusters and vagrant field stars. Here we find the oldest stellar populations with estimated ages of 12–14 billion years. The halo population of stars and star clusters is also unique in its chemical composition and kinematics. The outermost globular clusters have elemental abundances that are 1/100 that of the Sun and its stellar neighbors. They also exhibit the most disordered orbital trajectories. The inner globulars are a bit more metal rich and show a modicum of slow rotation. Some astronomers ascribe these differences to their respective origins. The outer globular clusters likely formed as part of separate dwarf galaxies that have since surrendered to the overwhelming gravity of the Milky Way. By contrast, the inner clusters may have been home-grown inside colliding parcels of gas—when the Milky Way looked more like a semi-spherical blob than today's disk-dominated system.

More clues to the Milky Way's early history can be found within the thin disk. Multiple measures of stellar age indicate that the disk came into being around 9–12 billion years ago. The faintest white dwarfs, for example, have ages of about 3–8 billion years based on estimates of the time it takes for them to cool to these low luminosities. Addition of the progenitor's stars' thermonuclear ages yields birthdates going back about 8–10 billion years. This reckoning should be regarded as a lower limit for the age of the thin disk, as it likely took some time for the thin disk to settle down and begin forming stars. Similarly, the oldest surviving star clusters in the thin disk yield minimal ages of 9–10 billion years. Star counting also shows a bump in the numbers of intermediate-mass stars whose total thermonuclear lifetimes are around 10–12 billion years, suggesting something special happened at this time to boost the numbers (see figure 4.4). Even the radioactive decay of certain isotopes observed in the spectra of old disk stars suggests formation times of 9–12 billion years.

All these measures show that the stars of the thin disk came into being a few billion years after the halo stars and clusters. Astronomers imagine a two-step formation process. First, large-scale gravitational collapse and condensation spawned the stars and clusters of the inner halo. After an uncertain number of giga-years, the remaining gas lost its kinetic energy through mutual collisions and radiative cooling. Compelled by its own gravitation, the gas collapsed into a disk whose plane was defined by the overall rotation of the proto-galactic system. Such a two-step scenario also helps to explain the strong disparity in elemental abundances between the halo and thin disk populations. Unlike the halo stars, the disk stars were beneficiaries of the chemical enrichment that the prior generations of stars provided.

Within the disk, we can perceive the full legacy of this chemical enrichment. Spectroscopic observations of Sun-like stars interior to the solar circle conclusively show up to three times higher iron abundances relative to the solar abundance. Beyond the solar orbit, the abundances drop to less than one-third solar. This prominent radial gradient in stellar iron abundances is also mirrored by the radial decline in nebular oxygen abundances as measured in HII regions. In both celestial venues, we are witness to the chemical legacy that has accumulated within the galactic disk over the eons.

Interior to the solar circle, the greater amounts of star-forming gas have led to enhanced stellar birthrates relative to those measured outside the solar circle. These elevated birthrates within the inner galaxy have accelerated the generation and distribution of heavy elements via supernova explosions. So, we are led to conclude that the galactic disk has been chemically evolving from the inside out. Because the radial abundance gradient is evident in long-lasting stars like the Sun, we can also infer that much of the chemical enrichment occurred early in the disk's history.

Whether the stars of the inner galaxy are characteristically older than those in the outer galaxy remains controversial. Observations of other spiral galaxies suggest that this is so, as the mean stellar colors appear redder in the inner portions of these galaxies—indicating greater proportions of older, redder stars. However, the redder stellar colors could also arise from higher abundances of heavy elements in the stars along with the reddening effects of dust in the inner disks of these galaxies, thus compromising the interpretation of greater age. To really ascertain the distribution of stellar ages throughout the disk of the Milky Way, astronomers will need to carry out dedicated spectroscopic surveys of billions of stars. These sorts of surveys are just beginning to be implemented.

Mysteries of the Bulge

Similarly ambiguous conclusions pertain to the galactic bulge, where a dizzying variety of stars can be found. Indeed, we don't really know when or how the bulge was built. If the bulge represents the most nucleated residue of the ancient halo, then it should contain lots of truly ancient stars. This appears to be the case, although with a twist. Many of the old stars exhibit in their spectra unusually high chemical abundances. Moreover, the abundances of oxygen and magnesium relative to that of iron appear unusually enhanced—telling scientists that the stars were recipients of matter from oxygen- and magnesium-rich supernovae rather than their iron-rich counterparts. The O- and Mg-rich supernovae came from short-lived massive stars (SNe of type II), whereas the Fe-rich supernovae are thought to have resulted from the obliteration of white dwarf stars in closely interacting binary systems (SNe of type Ia). This observed disparity in abundances tells us that the bulge stars were made after the births and deaths of the first massive stars but before the much later dissolution of intermediate-mass stars into white dwarfs. In other words, the stars of the classical bulge were forged early and quickly—before the type Ia supernovae had a chance to fortify them with iron. Whether these stars were formed as part of the bulge or were created elsewhere and subsequently transported to the bulge remains hotly contested.

Meanwhile, the stars of the inner bulge/bar appear to mimic the stars of the disk rather than those of the halo. Intermediate in metallicity, age, and rotational velocity, these stars bespeak episodes of wholesale migration from the surrounding disk into the Galaxy's bar-dominated inner sanctum. To be a star among the bulge/bar population is tantamount to being a flagrant vagrant, defying any sort of attempt to pin down its place and time of birth. Like Times Square in New York City, the Galaxy's inner bulge/bar is in a constant state of flux. Best to be wary of making any hasty conclusions.

Looking Out—Looking In

Somewhere in the Hubble deep fields there is a twin of the Milky Way as it was some 10–12 billion years ago. Our sundry guessing games regarding the Galaxy's origin and early history suggest that we wouldn't see too much. Perhaps we could perceive a starbursting proto-bulge of modest size (see figure 11.3). Hallmarks of the central starburst would include redshifted ultraviolet starlight and Lyman-alpha hydrogen line emission from the supergiant HII regions that then populated the proto-bulge. We might also find some evidence for a starbursting inner disk, whose rotation would induce recognizable Doppler shifts in any spectral-line emission that can be mapped. Some evidence for rotating disks at redshifts of 1–2 and corresponding lookback times of 7–9 billion years are already in-hand. These were obtained by observing redshifted H-alpha hydrogen line emission with adaptive optics techniques at the 8.2-m diameter Very Large Telescope in Chile.

Although we suspect a flurry of infalling dwarf galaxies during our Galaxy's formative years, these galactic building blocks may be too faint to detect unless they are vigorously starbursting. We may have better luck with the parcels of gas that would have been smashing into one another prior to the creation of the innermost globular clusters. The shock heating from these collisions could be sufficient to excite the molecular hydrogen in these clouds to glow brightly in the far-infrared—and with the subsequent redshifting—to be detected at radio wavelengths. Then again, we might be overwhelmed by brilliant quasar activity, as the newborn supermassive black hole feeds upon its immediate surroundings.

Through collective studies of luminous infrared galaxies at intermediate redshifts, astronomers have begun to perceive a possible sequencing in the phases that a galaxy like the Milky Way might undergo. Once enough primordial blobs have come together, the subsequent starbursts would create enormous quantities of dust. As the microscopic dust grains absorb the ultraviolet and visible starlight from the embedded starbursts, they would warm up and re-radiate the energy at mid- and far-infrared wavelengths. This phase has been observed in the form of Ultra-Luminous Infrared Galaxies (ULIRGs) out to redshifts of a few and lookback times in excess of 10 billion years. As the proto-bulge grows in girth and mass, the central black hole should also begin to take form. Once it attains a mass of several million Suns, the black hole would be hard to miss—shooting enormous jets through the surrounding bulge and photo-evaporating the ambient gas and dust. In a few billion years, the supermassive black hole would have managed to evacuate its immediate environment and so become directly observable as a quasar. We see this phase as a peak in quasar activity at redshifts of 2–3 and lookback times of 9–10 billion years.

Our search for the Milky Way's primordial twin may not be long in coming. New—more powerful—telescopes will soon address the so-called Dark Age, when galaxies were just taking form. Sometime after 2014, the James Webb Space Telescope (JWST) is slated to supplant the Hubble Space Telescope as the preeminent deep sky imager at visible and infrared wavelengths. Its 6.5-m diameter mirror will collect seven times more light than the HST—allowing it to peer out to redshifts of 5 or more and lookback times exceeding 12 billion years. Other telescopic facilities poised to raise the veil of the so-called Dark Age include the Herschel Space Observatory that operates at far-infrared and sub-millimeter wavelengths and the ground-based Atacama Large Millimeter Array (ALMA) in Chile that is optimized for millimeter-wavelength observations. ALMA, in particular, will be able to sense the first galaxian starbursts, and the dusty shrouds that they create, out to redshifts of 10 or more—when the universe was no more than a billion or so years old.

These and other new telescopes will also help us to diagnose the various legacies within the Milky Way that have not yet been erased by the passage of time. Innovative ground-based and space-borne telescopes equipped with powerful multiplexing spectrometers will soon harvest the spectra of billions of stars throughout the Galaxy. Each of these spectra contain within them vital clues to the star's age, chemical composition, and trajectory. Through intensive computer processing, astronomers hope to convert these spectroscopic clues into historical patterns of co-evolving and dispersing stellar groups. Much like a forensic scientist, they anticipate learning what happened where, when, and how throughout the Galaxy. Unlike their forensic counterparts, however, astronomers must contend with a crime scene that spans more than 100,000 light-years of galactic territory and involves more than 10 billion years of transformative history. A daunting challenge, indeed.

CHAPTER 12

. . • • . .

LIFE IN THE MILKY WAY

Lately, it occurs to me
What a long strange trip it's been.

— Robert Hunter and the Grateful Dead in "Truckin" (1970 CE)

So remember, when you're feeling very small and insecure,
How amazingly unlikely is your birth,
And pray that there's intelligent life somewhere up in space,
'cause there's bugger all down here on Earth.

— Monte Python in "The Meaning of Life" (1983 CE)

Fifteen hundred years ago everybody knew the Earth was the
center of the universe.
Five hundred years ago, everybody knew the Earth was flat,
and fifteen minutes ago, you knew that humans were alone on
this planet.
Imagine what you'll know tomorrow.

— Agent K in *Men in Black* (1997 CE)

IN THE LAST ELEVEN CHAPTERS, this book has presented the Milky Way in all its material diversity, spatial complexity, and temporal mutability. While writing these chapters, I have intentionally endeavored to portray our home galaxy as a living, breathing "organism"—one that emerged from the chaos of the Hot Big Bang some 12 billion years ago, and that is still very much alive with the fertile pyrotechnics of star-birth and star-death. What remains is an attempt to connect all these galactic theatrics to the emergence of our very selves—and perhaps to other life-forms that may be squiggling hither and yon elsewhere in the Galaxy.

As human beings on an increasingly crowded terrestrial stage, we have slowly begun to comprehend the importance of environmental factors in shaping and directing our lives. Today, we can no longer avoid addressing—and redressing—the many consequential connections that bind us to our physical, biological, and social surroundings. This sort of environmental thinking is most readily fostered within our own homes and hometowns, but can extend as far as the Galaxy itself.

Consider the African folk proverb that states "It takes a village to raise a child." Expanding upon that wise adage, I would contend that *it takes a galaxy* to cultivate stars, planets, and the conditions for complex life to emerge. This evolutionary process has played itself out to an uncanny degree on the moist surface of one rocky planet we know as Earth. In orbit around an unremarkable star, the Earth and the rest of the Solar System were forged after sufficient chemical evolution had proceeded in the Milky Way to foster a silicate planet with a magnetic iron core, a watery ocean, and an atmosphere suffused with nitrogen, carbon, and oxygen. Currently, the Solar System happens to be traipsing through a hot, rarified bubble that has been expanding within the disk of the Milky Way for the last 50 million years or so. The Local Bubble was likely inflated by one or more supernova explosions that detonated somewhere in the direction of Scorpius or Centaurus (see chapter 4). This now all-enveloping cavity provides visceral testimony to the impact of massive stars on galactic ecosystems throughout the disk. Indeed, the Earth owes its chemical pedigree to thousands of generations of massive stars that fervently lived and violently expired well before the Solar System's birth some 4.6 billion years ago.

Although we don't know where or how the Earth and Solar System formed, we can infer that the pre-solar nebula was part of a clustered system of stars which included a few massive members. Some of these brilliant behemoths must have died shortly before the Solar System's birth, because we can detect their rare isotopic effluvia today in meteorites. Unusual proportions of Magnesium-26 (the daughter of radioactive Aluminum-26), Nickel-60 (the daughter of Iron-60), and the "daughters" of other short-lived isotopes inform astrophysicists that one or more nearby supernova explosions injected the pre-solar nebula with newly forged heavy elements just before it congealed into planets. The precious metals that adorn our fingers and life-giving iron that reddens our blood may owe at least some of their origins to this last-minute episode of violent nucleosynthesis inside our galactic birthcloud.

In this last chapter, I consider the amazing cosmogenic and biochemical experiments that have given rise to life on Earth—and speculate upon other biotic adventures that may be running rampant throughout the Milky Way. Although scientists have yet to find evidence for life anywhere beyond Earth, that niggling absence of evidence has not stopped us from hypothesizing the existence of extraterrestrial life or from seeking it out. Indeed, the more we have learned, the more we are encouraged to think that simple microbial life is common, and that more complex and technologically advanced life-forms are at least possible. Perhaps we—or a subsequent Terran species—will make contact with another technologically communicative species on another world. In this spirit, I end the book with a *Galactic Manifesto* that calls us to develop and adopt an ethos worthy of communicating our history, science, art, and culture throughout the Milky Way.

Conditions for Life

What does it mean to be alive? To look at life on Earth is to be presented with a seeming chaos of efflorescence. No wonder anti-evolutionists invoke divine intervention to explain the amazing varieties and capabilities of species on Earth. What is missing is the timeline—a fantastically rich storybook that goes back to the first microbes some 3.7 billion years ago. In this primeval tale, we can perceive the basic essentials of life as it first emerged—and has since flourished—on Earth.

In its most rudimentary sense, life as we know it reduces to two functions—metabolizing and reproducing.[1] Motility may be extremely important to most life-forms, but it may not be necessary or sufficient to being "alive." Now, metabolism is a loaded word that implies many things to different people. From the perspective of a galactic astronomer such as myself, it seems fairly familiar in that it involves the transfer of matter and energy between the entity of interest (in this case, the life-form) and its environment. By that definition, a star could be regarded as a metabolizing entity! Clearly, there must be more to this business of metabolism. For biologists, metabolism involves a set of chemical reactions that modifies a molecule into another for storage, or for immediate use in another reaction, or as a by-product that remains or is expelled. These same scientists typically identify the pertinent metabolizing entity with the cell. Equipped with a permeable membrane of nitrogen-rich proteins and carbon-rich fatty acids along with all sorts of reactive goodies on the inside, the cell appears to provide the optimum solution for most metabolic activity.

Then there's reproduction. On Earth, this function is currently informed and implemented by deoxyribonucleic acid (DNA)—a twisted ladder of phosphates and hydrocarbon sugars that can break apart and clone itself from the ambient soup of chemicals that exist in the cell nucleus. DNA-mediated reproduction drives the other hallmarks of life on Earth—growth and evolution. Growth relies on cell replication, while evolution involves the two-step process of genetic mutation and natural selection (aka "survival of the fittest"). It is entirely possible that the earliest life was not nearly as adept at evolving in response to changes in its environment. If so, various versions of primitive life may have come and gone during the first few hundred million years until there came along a life-form that could transmogrify with the times.

DNA-encoded and driven life may not have been the only game in town either. Some biochemists have imagined a more primitive early life that relied on the simpler ribonucleic acid (RNA) molecule to carry out *both* functions of reproduction and metabolism. In addition to DNA and RNA, other nucleic acids have been put forward as alternative information-bearing molecules. Like DNA and RNA, they involve long chains upon which the encoding can proceed. They also share common ingredients including phosphorus (in almost all cases), amino acids full of nitrogen, and scads of hydrocarbons—all maneuvering within the favored solvent of water.

All of these ingredients can be found upon the galactic stage. Polycyclic aromatic hydrocarbons (PAHs) announce their presence at infrared wavelengths in settings as diverse as the atmospheres of red giant stars, the nebular disks of proto-planetary systems, and the myriad star-forming clouds that permeate our Galaxy's interstellar medium. Interstellar alcohols (methanol and ethanol) and even a rudimentary sugar (glycoaldehyde) have been discovered at radio wavelengths (see table 3.1). Nitrogen-rich amino acids don't glow as readily in nebular settings but are inferred from their identification in carbonaceous chondrites (carbon-rich stony meteorites) that have been found on Earth. Phosphorus is surprisingly rare on Earth, yet can be found at much higher concentrations in iron meteorites. Some scientists contend that our planet's paltry quota of this life-giving element was delivered primarily via meteoritic bombardments. Reactive sulfur is frequently belched up from the Earth's interior as part of the incessant churning of the Earth's upper mantle. Beyond Earth, sulfur makes up 2–7 percent of stony meteorites and dominates the volcanic chemistry of Jupiter's moon Io. Water has been identified in the most uncanny of places—as vapors in emission nebulae, ices in proto-planetary disks, and liquids below the icy crusts of the Jovian moons Europa, Ganymede, and Callisto. In 2005, the *Cassini* spacecraft imaged geysers of liquid water erupting from the surface of Saturn's moon Enceladus. The liquid form of water is especially important to biotic processes, as it provides an essential solvent for making the sundry hookups and energy transfers.

From the single precedent of our living Earth, we are left to conclude that life requires a wet, hydrocarbon-rich environment that is privy to oxygen and nitrogen, traces of other relatively heavy elements (specifically phosphorus, sulfur, sodium, magnesium, chlorine, potassium, calcium, and iron), along with reliable sources of free energy. It is entirely possible, however, that other biochemistries may prevail elsewhere. For example, alcohol, ammonia, and methane can remain in liquid form at much lower temperatures than water. Therefore, they could conceivably serve as solvents in which simple elements and molecules would react to form new kinds of complex molecules. Some cosmochemists have begun to explore such chemistries occurring in the methane seas that have been recently confirmed on Saturn's largest moon Titan. Cosmochemists have also considered replacing carbon's pivotal role in the chemistry of life with some other key element—such as sulfur or even silicon. These alternative chemistries have been less successful, however, as it is difficult to outdo carbon's chemically promiscuous ways.

Making Life

We know that life emerged on Earth shortly after it had cooled from a molten state, roughly 4 billion years ago. Some scientists fix the date at 3.8 billion years ago, immediately following the cessation of the Late Heavy Bombardment (whose impact craters and basins can still be seen on the Moon). Alas, we may never know *how* life's emergence came about. Although the physical circumstances of early Earth are fairly well delineated, the chemical processes on land, in the atmosphere, and in the oceans are far less constrained. Indeed, we are first faced with trying to understand how Earth acquired its precious store of liquid water.

Water's Origins

Despite compelling evidence for water throughout our Milky Way Galaxy and within our local Solar System, the specific origin of water on Earth remains controversial.[2] We know that the Earth formed some 4.6 billion years ago, as the primordial Solar System gravitationally congealed from the pre-planetary disk of debris that surrounded the proto-Sun. Once the Sun "turned on" its thermonuclear fires, things began to change in a big way.

Some scientists contend that the inner Solar System was too hot for any water to remain in the rocky bits that ultimately came together to form the Earth. They look to wayward comets and asteroids that formed farther from the scalding Sun as the key provisioners of Earth's oceans. However, others find important discrepancies in the relative amounts of regular water and "heavy" (deuterated) water in their comparisons of the Earth, comets, and meteorites. The few comets that have been chemically probed appear to have far more deuterated water than Earth's oceans—indicating (to some scientists) that comets could not have delivered the bulk of Earth's water.

Meteorites are interplanetary rocks of various kinds that have fallen to Earth. They are thought to represent pieces of much larger asteroids that formed beyond Earth's orbit—between the orbits of Mars and Jupiter. Some meteorites are found to be rich in water. Others are bone dry. Again, chemical and isotopic analyses rule out asteroidal meteorites delivering most of Earth's water, as they find that the early Earth was likely the beneficiary of mostly dry meteorites. The only way to wiggle around this situation is to have a single, unusually wet, and very large (Moon-size) body to hit Earth shortly after its birth. Something like the Jovian moon Europa could have done the trick.

Perhaps surprisingly, several scientists have gone back to square one—working out scenarios which retain the water in the inner Solar System despite the intense solar heating. Here, the water was in the form of a hot vapor that stuck to tiny grains of rock which then aggregated to build up pebbles—and ultimately—a wet Earth. Once the Earth's crust began to solidify, it would have belched out huge quantities of water during a period of rampant volcanic eruptions. As the saturated atmosphere cooled, it began to rain—and rain—and rain, filling the Earth's basins with the water that we inherit to this day.

Research on water's terrestrial origins will likely benefit from future interplanetary probes, whereby the isotopic composition of water across the Solar System can be assayed. That will help to pin down the status of oceanic water on Earth relative to the larger context of planetary, asteroidal, and cometary sources. Meanwhile, sub-millimeter spectroscopic observations of icy objects in the Kuiper Belt may help to determine the isotopic composition of these pristine Solar System objects beyond the orbit of Neptune. Closer to home, deep-sea probes will help to determine the isotopic composition of water emerging from the upper mantle.

Origins of Biotic Molecules

Since the 1950s, laboratory chemists have been brewing mixes of gases and liquids that were thought to characterize the primitive Earth. By heating the mixtures to temperatures similar to those in volcanic vents or by zapping them with currents of electricity (simulating lightning), they have succeeded in concocting all sorts of fascinating goop. Chemical analyses of the residues have revealed amino acids, fatty acids, sugars, and other complex molecules that are known to play important roles in building proteins, nucleic acids, and other key components of living matter.

By dissolving particular fatty acids into water, some biochemists have even been able to replicate rudimentary cells. The fatty "lipid" molecules have water-loving heads and water-repelling tails which cause them to self-assemble into hollow spheroids. These fatty bags arising in solution demonstrate that the early Earth may have hosted similar spontaneous constructions, inside of which unique chemistries and metabolic processes ultimately ensued.

Although chemists have yet to build RNA and DNA from scratch, they have come tantalizingly close. One particular challenge is to understand the dominant helicity that is found in these and many other biotic molecules. Somehow, life on Earth seems to prefer molecules with common twists, turns, and other asymmetries—what scientists term "homochirality." For example, all of the amino acids that make up proteins on Earth are "left-handed," while all of the sugars that form the backbones of DNA are "right-handed." This preference has suggested to geochemists that the first biotic molecules were built on some sort of mineral template. The crystal surfaces of pyrite, quartz, feldspar, and even grains of clay have been invoked to provide these sorts of templates, as they favor chemical bonds with particular orientations. Other chemists have shown that neat stacks of PAH molecules could have provided the "key" for growing nucleic acids with the right attitude.

For now, there appear to be several promising pathways toward homochirality, but no clear winner.

Another challenge is to understand why only twenty amino acids out of a hundred possible variants are represented in the chemistry of life. It's as if terrestrial life had made a choice to use only those amino acids that could interact with DNA and RNA to produce the proteins that underlie all forms of life on Earth. Was this a random occurrence or was there some sort of logic involved? If random, we would have to seriously consider the possibility that extraterrestrial life-forms could be based on an entirely different set of amino acids and with a whole other suite of nucleic acids and proteins as the result.

Life on Earth

Like water's uncertain origins, the first life on Earth remains elusive. Was it cooked up out of preexisting organic molecules that had concentrated in evaporating pools? Did it arise from mineral-rich volcanic vents? Or were the essential seeds of life delivered to Earth by comets and meteorites? However life jumpstarted on Earth, the geological record now shows that it quickly enveloped the planet's surface (see figure 12.1). By scrutinizing layers of sea-floor sediments, scientists can tell that the seas were teeming with microbial life less than a few hundred million years following the emergence of the very first cellular life-forms roughly 3.5 billion years ago. These microbes were most likely akin to the cyanobacteria that still flourish in tide pools and freshwater ponds today. Within the next billion years, humongous colonies of the photosynthesizing bacteria began to characterize the seashores. Known as stromatolites, these bacterial mats grew to thicknesses of many meters. The first single-celled life with a nucleus (eukaryotes) appeared roughly 2.5 billion years ago—more than a billion years after the first cellular life. Multicellular organisms began to leave fossil evidence another billion years later—about 1.5 billion years ago. Yet another 900 million years would pass before the emergence of worms, jellyfish, mollusks, and other relatively complex animals.

As photosynthesizing organisms overran the shallows, they pumped ever increasing amounts of free oxygen into the Earth's oceans and atmosphere. This oxygenic infusion is thought by some geochemists to explain the banded iron formations that were laid down between 3 and 2 billion years ago. In this scenario of geological origin, the dissolved oxygen in the existing seas combined with iron atoms in solution—producing extensive sediments of iron oxide (rust) that accumulated on the seafloors. Other geochemists have proposed that microbes metabolized the oceanic iron, and so were responsible for precipitating out the banded deposits of iron oxides . . . even before the oceanic oxygen levels had ramped up. Sometime around 500 million years ago, the oxygen levels crossed a critical threshold coincident with a huge increase in the variety of animals that are recorded in the fossil record. Scientists have debated the meaning of the so-called Cambrian Explosion. Did it involve a major amplification in the number of body types (phyla), or was it more an increase in the *sizes* of preexisting species that led to greater numbers of creatures surviving the fossilization process? Recent evidence indicates that pre-Cambrian life was in fact much less diverse, and that the Cambrian Explosion truly represents a dramatic radiation of phyla and corresponding species.

Whatever the Cambrian Explosion involved, it marked the epoch when life itself began to dictate the form and content of Earth's surface. Take the Alps, for example. Much of this massive uplifted landform is comprised of microbial fossils—the carbonate sediments of countless tiny sea creatures. The Earth's atmosphere shows the effects of life even more clearly. Its current quota of 21 percent free oxygen can hardly be sustained unless new life is continuously injecting fresh oxygen to replace that which gets bound up into rocks and seawater. On both the land and in the sea, we can see that terrestrial life has created an amazing state of disequilibrium which continues to evolve up through the present day. Humans are just the latest to add to the litany of environmental impacts. Although we have made our mark on the Earth's atmospheric content—and perhaps on the evolving climate—we continue to be dwarfed by the effects of far more primitive and durable life-forms.

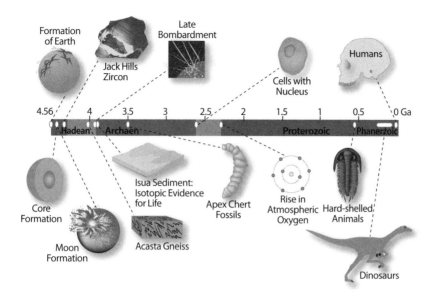

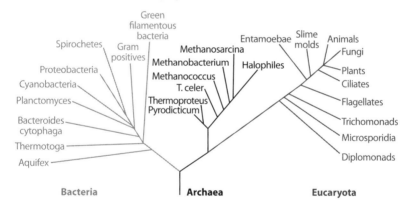

Phylogenic Tree of Life

FIGURE 12.1. (*Top*) Schematic history of life on the Earth—from its formation 4.6 billion years ago through the first 3 billion years of single-celled life to the emergence of animals and humans. (*Bottom*) The updated "tree of life" shows three main branches, each of which is dominated by single-celled microbes. The Archaea have no nuclei, with other properties that distinguish themselves from the more familiar Bacteria. They are thought to have emerged first, perhaps as thermophilic (heat-loving) extremophiles. The Eukaryotes contain nuclei and include multicelled organisms, such as fungi, plants, and animals. [(*Top*) Illustration courtesy of Andrée Valley (http://andreevalley.com). (*Bottom*) C. Woese, NASA Astrobiology Institute; see http://www.uwyo.edu/dbmcd/molmark/lect11/tree.jpg)]

Extremophiles can survive in environments that would be horrifically hostile to most "normal" life such as you, me—and even the microscopic mites nestling in our respective eyelashes. These hardy organisms survive—and often thrive—in conditions of extreme heat, cold, pressure, acidity, salinity, and radioactivity. Extremophiles are not just a curiosity, however, as they may currently comprise the bulk of the biomass on (and within) Earth. Moreover, they may have dominated the earliest epochs of terrestrial life, when high temperatures, pressures, and radiation levels were more the norm than the exception (see figure 12.1). For astrobiologists, the study of extremophiles provides an essential window to pondering the sorts of microbial life that may exist below the frozen surface of Europa or in the nitrogen- and methane-rich seas of Titan.

Most extremophiles are fairly simple—consisting of single cells or of cellular colonies. The slippery dark scum that lives on coastal rocks within the salty intertidal "splash zone" is a fairly common example of a halophile, or salt-loving extremophile. Here, the diurnal wetting and drying leads to large buildups of salt that could not be tolerated by the more submerged seaweeds. More complex but common examples of extremophiles include the many kinds of lichens that manage to grow on bare rocks and within arctic snowpacks. Consisting of symbiotic fungi that ingest water and minerals together with foliate algae that provide energy through photosynthesis, the lichens demonstrate how daunting environmental challenges can lead to very clever evolutionary solutions.

Beyond these relatively familiar examples, extremophiles on Earth include the vividly colored thermophiles that inhabit the near-boiling hot springs in Yellowstone National Park, the psychrophiles that live on microscopic grains of rock deep inside antarctic icepacks, the endoliths that have been found lining bore holes drilled 5 kilometers beneath the Earth's surface, and the radiation-, heat-, acid-, and dehydration-tolerant *deinoccocus radiodurans* bacteria that have been discovered thriving inside the radioactive cores of nuclear reactors.

Then there are the black smokers. Deep beneath the Earth's oceans—along the volcanic ridges that wend their way around the

Earth like the stitches of a baseball—innumerable rock "chimneys" have been found venting superheated water along with lots of minerals that color the water black. Despite their pressurized, scalding, and totally obscured conditions, the so-called black smokers serve as energizing hosts to all sorts of microbial life. Scientists think that this life feeds on sulfurous compounds and the upwelling heat as proxies for the sunlight that is sorely lacking at these depths. The microbes, in turn, nourish entire ecosystems of more complex organisms—including shrimps, crabs, and fish. Most amazing are the giant tubeworms—strange animals that can grow to lengths of 3 meters yet have no mouth, eyes, or stomach. They survive through a symbiotic relationship with the billions of bacteria that live inside of them.

Could there be comparable food chains based on sources of chemical energy beneath Europa's frozen crust? Before discovering the black smokers beneath our own oceans and the elaborate ecosystems associated with them, we would have never imagined this possibility. Because of these unexpected discoveries, our concept of a planetary system's "habitable zone" now extends well beyond that annulus where the central star can irradiate and so maintain a planet's surface water in liquid form.

Life beyond Earth

The emerging field of *astrobiology* got its first major boost in 1961, when ten radio technicians, astronomers, and biologists convened in Green Bank, West Virginia, to address the prospects for intelligent, communicative life beyond Earth. In preparation for the meeting, Frank Drake of the National Radio Astronomy Observatory (NRAO) developed an equation that formulates the sundry (im)probabilities that combine to yield an estimate for the number of telecommunicating planets in the Milky Way.[3] Beginning with the total number of stars in our Galaxy, the eponymous "Drake Equation" lays down a gauntlet of challenging questions—some of which are still as uncertain as ever.

The equation itself has a disarming simplicity, consisting of just seven factors in its most succinct form:

$$N(\text{comm}) = N(\text{star}) \times f(\text{sun}) \times f(\text{plan}) \times n(\text{hab})$$
$$\times f(\text{life}) \times f(\text{int}) \times f(\text{comm}),$$

where:

$N(\text{comm})$ = number of intelligent, communicative civilizations in the Milky Way,
$N(\text{star})$ = number of stars in the Milky Way,
$f(\text{sun})$ = fraction of stars similar to the Sun,
$f(\text{plan})$ = fraction of Sun-like stars with planets,
$n(\text{hab})$ = number of habitable planets per Sun-like planetary system,
$f(\text{life})$ = fraction of habitable planets with life,
$f(\text{int})$ = fraction of inhabited planets with intelligent beings,
$f(\text{comm})$ = fraction of planets with intelligent species that are technologically communicative.

To see how these cofactors collectively winnow the prospects, let's consider each factor in its turn, and propagate the results. As you will see, the first two factors are fairly well established, the third factor will soon be pinned down, and the rest remain topics for conversations over cocktails.

N(STAR) = NUMBER OF STARS IN THE MILKY WAY ($\sim 100 \times 10^9$)

Although an actual count of all the stars in the Milky Way is impossible, we can obtain a near-complete census of the stars within 15 light-years of the Sun. The resulting density of 0.1 stars/pc³ (or 2.9×10^{-3} stars/ly³), when combined with models of the overall distribution of stars in the Milky Way, yields a total extrapolated number of about 100–400 billion stars. These results are consistent with observations of the total starlight from other nearby giant spiral galaxies.

F(SUN) = FRACTION OF STARS SIMILAR TO THE SUN (~ 0.1)

To enable the evolution of life, a stable, long-lasting star is preferred. Stars significantly more massive than the Sun are too short-lived for life to take hold, while stars of much lower mass are prone to violent flaring. For these reasons, stars of spectral type F,

G, and K make the most promising candidates. The census of stars in the solar neighborhood suggests a fraction of 0.2, or more conservatively, 0.1. That leaves about 10 billion Sun-like stars. Many exoplanetary scientists are revisiting the cooler and dimmer stars—despite their flaring ways—as submerged, buried, or even sheltered life-forms might be able to survive these outbursts. Therefore, the quoted estimate of 10 billion viable stellar hosts is surely on the conservative side.

F(PLAN) = FRACTION OF SUN-LIKE STARS WITH PLANETS (~0.1)

Two pieces of evidence indicate that planetary systems are fairly common. One is the fraction of stars in binary systems. Roughly 75 percent of all the stars are parts of binary or multiple star systems. These stellar bunchings can be dynamically hostile to any associated planetary systems. The remaining 25 percent of stars are thought to be more likely hosts of planets in stable orbits. The second piece of evidence is the actual harvest of planetary systems that has been obtained from tracking the Doppler-shifted light of nearby stars—and more recently, the transits of planets in front of their hosting stars. Of the thousands of stars that have been surveyed to date, more than 700 appear to be in the gravitating presence of planets. These radial-velocity and transit-based surveys are biased toward the detection of massive planets that are very close to their host stars (see chapter 4). However, they provide a reasonable lower limit of 10 percent for the fraction of stars with planets, leaving about 1 billion Sun-like stars. We should soon get much better statistics on this fraction, as the *Kepler* space mission and several dedicated ground-based programs complete their far more extensive searches for exoplanetary systems.

N(HAB) = NUMBER OF HABITABLE PLANETS PER
SUN-LIKE PLANETARY SYSTEM (~0.1)

Here, we begin to run into speculative territory. Current observing technology is insufficient to detect planets that are significantly less massive than Jupiter, i.e., Earth. Therefore, we are left with extrapolations from our own situation. Of the eight major planets in the Solar System, only Earth occupies the so-called water zone, the annular region whose distance from the Sun allows surface water to be in liquid state. Earth barely qualifies, the greenhouse effect of its atmosphere providing enough additional heating to keep the surface from freezing over—most times. Beyond the water zone, the subsurfaces of Mars, Europa, Ganymede, Callisto, and Enceladus may also harbor liquid water—wherein sources of chemical energy can substitute for the Sun's warming rays. The biochemistry of life on Earth relies on water as the primary chemical solvent. Other chemicals, such as ammonia, may qualify as solvents in other, more frigid worlds. Lacking further insights, our "water zone" requirement suggests a number of one to four habitable worlds per system, or more conservatively one habitable planet every ten systems. We are then left with about 100 million habitable planets in the Milky Way.

F(LIFE) = FRACTION OF HABITABLE PLANETS WHERE LIFE ARISES (~0.1)

Without any hard evidence yet for life of any kind beyond Earth, astrobiologists are still very optimistic about the prospects for life evolving on habitable planets elsewhere in the Milky Way. As we have seen earlier in this chapter, the ubiquity of complex organic molecules in star-forming regions and of amino acids in meteorites tells us that the necessary ingredients are available. Moreover, the uncanny flourishing of microbial life in the most extreme terrestrial environments attests to the tenacious qualities of single-celled life-forms. A naive consideration of the terrestrial record reveals a living history that goes back at least 3.5 billion years, amounting to more than 60 percent of the 4.6 billion-year age of Earth. If we shamelessly extrapolate from our own situation, then we might guess that one in ten habitable planets actually sports life of some kind. Our educated—and seemingly conservative—guess would yield about 10 million life-bearing planets in the Milky Way.

F(INT) = FRACTION OF INHABITED PLANETS WITH INTELLIGENT BEINGS (10^{-3})

Once again, our lack of information beyond Earth leaves us grasping at our own reflection. Using relative lifetimes as our guide, we find that human-like intelligence has marked the Earth for only a few million years, or about 1/1,000 the history of life on Earth.

Perhaps intelligent life-forms will continue to prevail on Earth for eons to come—thus upping the extrapolated odds. For now, however, a fraction of $\sim 10^{-3}$ is as good a guess as any. Our tally of the Milky Way is now reduced to about 10,000 planets with intelligent life.

F(COMM) = FRACTION OF PLANETS WITH INTELLIGENT SPECIES THAT ARE TECHNOLOGICALLY COMMUNICATIVE ($\sim 10^{-4}$)

Here be dragons. This fraction critically depends on how long a technologically communicative species can functionally survive. To date, our species has been transmitting radio energy for about 10^2 years, or roughly 10^{-4} of our history as hominids. Perhaps we can keep this up for another thousand years or more—who knows! Meanwhile our guessing game would lead to the following total number of telecommunicating planets in the Milky Way: N(comm) ~ 1. We may be it!

As the previous example has shown, the Drake Equation provides lots and lots of wiggle room for scientists and nonscientists alike to pontificate on the status of life in the Milky Way. Some scientists have made persuasive cases for microbial life being fairly ubiquitous while putting the kabosh on more advanced life-forms such as the plants and animals that dominate our myopic view of the living Earth. Others warn against our inevitable referencing to what we have found on Earth. Perhaps life has found other ways to permeate the Galaxy and evolve. For now, however, we are left marveling at our Earth-bound extremophiles and imagining what else might be out there.

Conditions for Conscious Communicative Life

While the Drake Equation offers a framework for thinking about the sundry cofactors that together determine the likelihood for conscious technologically communicative life, it does not address the more anecdotal aspects of fostering and finding such advanced life-forms. What kinds of conditions are really necessary to breed the likes of ourselves—or of even more capable organisms? Here, science fiction writers have provided important contributions to this admittedly speculative discussion. In his *Foundations Series,* Isaac Asimov imagined a Galaxy-wide empire on the cusp of collapse. The past, current, and future machinations of this empire were understood according to the tenets of "psychohistory," whereby larger populations of people yielded more predictable and coordinated behaviors.

Building on this idea, perhaps we should be focusing our attention on populous clusters of stars, where the potential for inter-cluster communication and cooperation would be greater. I would recommend targeting the open star cluster Messier 67 in the constellation of Cancer the Crab. Located a "mere" 2,700 light-years from us, M67 was formed approximately 4 billion years ago. That means the cluster's 500 remaining stars all have ages and chemical abundances similar to that of the Sun. Perhaps in M67, we may eventually find evidence for multiple worlds in coordinated communication.

Full-time cosmologist and part-time sci-fi author Fred Hoyle speculated on the sorts of cognizant life that might emerge within a dark dusty cloud of gas. In *The Black Cloud,* he portrayed such intelligent life as consisting of the cloud itself. Vast and amorphous, the nebular being excelled in harnessing radio emissions for interstellar communication. Although such diffuse life is well beyond our ken, we can take away from this tale the notion that all planetary systems form within dark molecular clouds, and that some of them may spend up to a hundred million years inside these murky realms. Blinded by the obscuring dust in these clouds, any advanced life-forms would have to rely on radio technologies to learn about and interact with inhabitants of other planetary systems. Consider the pseudo-ring of molecular clouds that has been found halfway to the galactic center (see figure 3.5). Within this annulus, the resident times of planetary systems inside molecular clouds must be significantly greater. If true, Hoyle's idea of a radio-biased intelligent life form becomes ever more plausible.

Then there is the "Goldilocks Hypothesis" for predicting where in the Galaxy intelligent life is most likely to emerge. Here, astrophysicists have delineated a galactic Habitable Zone that is favorable to the development of life. It is based on the finding that the abundance of elements heavier than helium varies significantly with galactocentric radius. Regions well beyond the solar orbit are relatively depleted in these heavier elements. That means the formation of rocky planets like Earth becomes suppressed. If life is in fact a planetary phenomenon where liquid water and rocky surfaces must intermingle, then the outskirts of the Milky Way may host too few rocky planets for intelligent life to emerge. Looking inward toward the galactic nucleus, we find the heavy-metal abundances approaching toxic levels. Moreover, we are also considering real estate that is dangerously close to lots of supernova activity along with the (currently) dormant supermassive black hole at the core. Perhaps that would not be such a good place to set up the business of life. By considering these various environmental factors, astrophysicists have estimated that the favored annulus for advancing life in the Milky Way is currently 22–29 thousand light-years from the galactic center—and that it is slowly widening with time.

This notion of dividing the Galaxy into fertile and barren sectors may depend on a wide variety of factors—including the local metal abundance, the position above and below the galactic midplane, the radially dependent frequency of spiral arm crossings, the proximity to recurrent starburst activity, and the spatially irregular history of impacts from infalling clouds. Then again, it might be life itself that will determine the character of the animate Milky Way. As the emergence of photosynthesizing life completely altered the environment and all subsequent life on Earth, so too the proliferation of life beyond Earth may similarly depend on what has gone before.

In this spirit, I cannot help but invoke the living galaxy that Gene Roddenberry envisioned in his original *Star Trek* television series, and that has since provided the context for all subsequent *Star Trek* series and movies. Here, the Milky Way is divided into the storied Federation, Romulan, Klingon, Cardassian, Kazon, Dominion, and Borg sectors (see figure 12.2). How could such decidedly different civilizations take such firm hold of the Galaxy? Well,

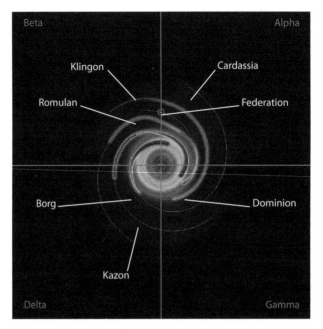

FIGURE 12.2. The civilizations in the *Star Trek* franchise of television shows and movies overlaid on a scientifically based rendering of the Milky Way Galaxy. These civilizations were established with the assistance of superluminal travel. Viral propagation of life-forms in the real Galaxy would take much longer, but could be realized on timescales of a billion or so years. [Courtesy of the Canadian Galactic Plane Survey science team (see http://www.ras.ucalgary.ca/CGPS/where/plan/)]

for one thing, they did not have to settle for traveling at velocities less than the speed of light. Boosted by their hyper-speed capabilities, the characters of *Star Trek* could hurtle themselves clear across the Galaxy well within the timespan of a single television episode. If we restrict ourselves to the scientific laws as we currently know them, however, such superluminal travel would be out of the question. Yet, all is not lost, as we have lots of time to mix things up on the galactic stage. Considering the 12 billion-year history of our home galaxy, it is entirely possible for a particular brand of life to propagate versions of itself over large expanses of the Milky Way in less than a billion years. The velocity dispersions among the various components in the disk are indeed sufficiently large for this sort of viral spreading to accrue. What technology cannot do, the Galaxy itself may enable.

Toward a Galactic Manifesto

In many ways, we have already begun to "infect" our little corner of the Milky Way. The various spacecraft that landed (or crashed) on the Moon, Mars, Titan, and Comet Tempel carried with them rich cargos of Terran bacteria. These life-forms have most likely died and dried out, but their genetic blueprints remain—ready for rehydration. Some of the genetic code may have been lost in translation, the result of cosmic-ray impacts beyond the protective magnetosphere of Earth. However, it is conceivable that significant portions of the bacterial genotypes have survived the onslaught of radiation—especially those that have ended up far from the Sun or the radiation belts of Jupiter. The same sorts of vestigial life are traveling aboard the robotic probes that have flown by or are still orbiting the many worlds in our Solar System.

Between 1977 and 1989, the *Voyager 1* and *Voyager 2* spacecraft flew past Jupiter, Saturn, Uranus, and Neptune—providing us with images that have forever changed our perceptions of the outer Solar System. The *Voyager* spacecraft are currently in the process of crossing the border that separates the solar wind from the surrounding interstellar medium—the so-called heliopause. Once beyond this bubble-like interface, they will become our first true galactic ambassadors. Besides transporting a veritable ark of bacterial corpses, these pioneering probes also sport gold-plated 12-inch records on which are inscribed the sights and sounds of Earth. The sounds span a wide variety of natural and artificial sources—including earthquakes, volcanoes, mudpots, oceanic surf, gentle breezes, violent storms, crickets, frogs, birds, elephants, whales, humans, tractors, trains, jets, a *Saturn 5* rocket launch, and a pulsar. Among the human sounds are greetings in fifty-four languages and ninety minutes of music from around the world. The spacecraft are not aimed anywhere in particular, but perhaps someday they will intersect other planetary systems or be intercepted by other sentient life-forms.

While the *Voyager 1* and *2* spacecraft are truly shots in the dark, our radio and television broadcasts are more like indiscriminate shouts into the void. Radio broadcasting began in 1906, and powerful radio and radar technologies were developed during World War II. Our resulting "radiosphere" has a radius of at least 60 light-years, inside of which are thousands of stars and perhaps hundreds of planetary systems. Television broadcasting began in the late 1920s and has been filling the airwaves with newscasts, sports events, sitcoms, and commercials ever since the 1940s. Carl Sagan, in his book *Contact,* portrayed radio astronomers receiving their first coded communication from an extraterrestrial civilization. It turned out to contain a television rebroadcast of the 1936 Olympic Games, with Adolf Hitler presiding over the opening ceremonies. In many ways, Sagan was onto something. We have been inadvertently communicating with the rest of the solar neighborhood for quite a while, with consequences that are as unpredictable as the content of our radio and television programming. Perhaps we should think about how we could do a better job of this.

The idea that we have already begun to be players on the galactic stage is pretty unnerving. How should we comport ourselves? And how long must we wait before we receive confirmation that somebody is paying attention? To answer the first question, I contend that we should begin to exercise our natural birthright as full-fledged citizens of the Galaxy. That means taking greater care in our mutual relations, so that other more advanced civilizations might regard us with some measure of respect. It also means deliberately broadcasting information on our history, science, art, and culture in ways that would represent us well to another equivalently advanced civilization. Call it interstellar marketing.

We might wish to aim these broadcasts toward specific stars or star clusters—as occurred in 1974 when the giant radio dish in Arecibo, Puerto Rico was used to broadcast a carefully coded message toward the globular star cluster M13 (see figure 12.3). Such directed communications toward specific targets increase the

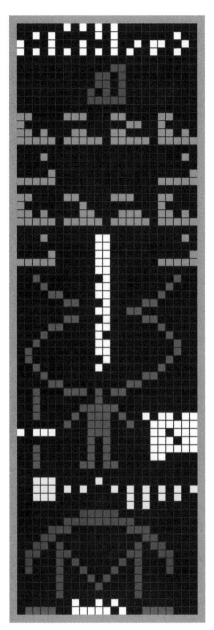

FIGURE 12.3. The coded message that the giant Arecibo radio telescope transmitted toward the M13 globular cluster of stars in 1974. Developed by Frank Drake, Carl Sagan, and others, the binary message can be decoded to reveal (*from top to bottom*) the numbers 1–10, the atomic numbers of hydrogen, carbon, nitrogen, oxygen, and phosphorus which together comprise deoxyribonucleic acid (DNA), the formulas for sugars and bases in the nucleotides of DNA, the numbers of nucleotides in DNA along with a graphic of DNA's double helix structure, a graphic representation of a human with coded information on the human's height and the number of humans on Earth, a graphic of the Solar System, and a graphic of the Arecibo radio telescope with coded information on the diameter of the transmitting dish. Because M13 is approximately 25,000 light-years away, we would have to wait at least 50,000 years (2,000 human generations) to receive any sort of reply. [Drawing by Arne Nordman, Wikipedia Commons, with permission via the Creative Commons Attribution-Share Alike 3.0 Unported License (see http://en.wikipedia.org/wiki/Arecibo_message)]

chances that our intended messages supercede the diluted chatter from countless other broadcasts. Once again, I would recommend targeting M67 and other relatively nearby clusters of similar age as our Solar System. We might also want to think about what sorts of questions we should be posing to our galactic neighbors. All of these considerations would need some sort of global forum for further deliberation. To be fair, I think that a special committee on Interstellar Communications would be best established within the United Nations—perhaps as part of the United Nations Educational, Scientific, and Cultural Organization (UNESCO). This could build on the impressive cooperation in crafting communications protocols that has already been forged by the SETI Institute, SETI League, Planetary Society, International Astronomical Union, and International Academy of Astronautics.

Regarding the second question on how long we will have to wait for a positive reply, I have no idea when (or if) we will ever receive an intelligible communication from another civilization— let alone one that responds to anything we have sent out. Moreover, I worry about the pertinent timescales. If intelligent life is sparse, then our signals will have to travel farther. Issues of powering such far-reaching communications become problematic. Also, the light-travel times begin to exceed our likely lifespan as a species. Ten years may seem a long time to wait, but time spans of 100, 1,000, 10,000, and up to 100,000 years may be more pertinent to the actual situation. If interstellar communication in fact requires 10,000 to 100,000 years to actualize, then we have to think about how we can propagate our communications efforts to the next version of ourselves—or to the next fully cognizant lifeform that emerges on Earth. This seems like an impossible task, but perhaps our initial forays at sending informative records and plaques aboard interplanetary probes can guide us.

What I'm recommending is nothing less than interspecies communications on Earth over evolutionary timescales. Consider what we know of the earliest Homo sapiens. Their existence over the past 200,000 years has been preserved in the primitive stone tools that they used and via the middens of seashells and animal bones that they accumulated. Now consider what our own civilization will be leaving behind. Of all the enduring constructs of stone and steel that we have created, precious few have been specifically designed to communicate information about ourselves beyond a few dozen generations. The stone and turf "passage tomb" of Newgrange in Ireland, the standing stones of Stonehenge in England, the pharaonic pyramids in Egypt, and the translational Rosetta Stone of later Egyptian provenance come closest to achieving this ideal—providing tangible keys to our cultures and cosmologies that go back 5,000 years. We can do much better than these precedents, however. By inscribing carefully encoded messages upon durable stone or metal surfaces—and situating these artifacts wherever we have sent out or attempted to receive interstellar broadcasts—we can provide salient clues regarding our own efforts to communicate beyond Earth. If, through advances in nanotechnology, we develop sufficiently long-lasting data storage devices, these can be embedded as part of the lower-tech cartouches, stelae, and monuments that have served us so well in the past.

In the first film version of Pierre Boule's novel *Planet of the Apes,* the only human edifice to survive the millennia was the Statue of Liberty—its uppermost parts jutting out of the coastal sands that had buried the rest of New York City. This fiction underscores how little we can predict what will remain of us after our demise. The periodic glaciations and sporadic super-volcanos on Earth are too adept at modifying the living landscape. That is why we may stand a better chance of being rediscovered by placing our calling cards on the surfaces of the Moon and Mars—where geophysical activity has long since ended and the erosive effects of micrometeorites take billions of years to accrue. Something like the "Monolith" in *2001: A Space Odyssey* might work, though with much clearer messaging. Indeed, the prospect of bridging the communications gap between generations—and species—may provide one of the most compelling justifications for NASA to develop a viable human presence on these worlds.

So, this book ends with a call to take on our rightful place as fully engaged citizens of the Galaxy. Over the past 500 years, we have looked skyward with ever greater capability, and in so doing, have expanded our perception of place—from the Earth to the

GLOSSARY

· · • • · ·

absolute luminosity   The radiant power of an object measured as the
  amount of energy emitted per unit of time. Commonly used units are
  ergs per second, joules per second (Watts), and solar luminosities,
  where 1 solar luminosity equals 3.84×10^{33} ergs/second or 3.84×10^{26}
  Watts.

absolute magnitude   The luminosity of an object measured in the logarithmically compressed units of magnitudes (M) and defined as the
  apparent magnitude (m) the source would have if located at a distance
  of 10 parsecs (32.6 light-years). The Sun has an absolute visual magnitude of $M_V = +4.83$, while Vega has an absolute visual magnitude of
  $M_V = 0.0$.

absorption spectrum   A spectrum of an object in which cool gases have
  absorbed light of certain wavelengths, producing gaps in the brightness
  at these particular wavelengths.

accretion disk   A disk of gas and dust that forms around a protostar as it
  settles down to become a pre-main-sequence star. Ambient material
  falls onto the accretion disk and spirals inward toward the protostar.
  On larger scales, active galactic nuclei are thought to have accretion
  disks surrounding them.

active galactic nuclei (AGNs)   Nuclei found in galaxies that exhibit unusually energetic behavior in the centers.

amino acid   Organic molecules containing carbon, hydrogen, oxygen,
  and nitrogen that serve as the building blocks of all proteins. Living
  matter is comprised of twenty different amino acids.

angstrom   A unit of length used in spectroscopy, equal to 10^{-10} meters.

antimatter   A subatomic form of matter that is of opposite charge to that of ordinary matter. Examples of antimatter include antiprotons, anti-electrons (positrons), antineutrinos, and antiquarks.

apparent magnitude   The brightness of an object as seen from the Earth, measured in units of magnitudes (m). The brightest stars in the sky are of first magnitude or lower; the faintest naked-eye stars are of magnitude m = 6. The most powerful telescopes can reach magnitudes as faint as m = 30.

arcsecond (arcsec)   A second of arc, a measure of angular extent. One arcsecond equals 1/3,600 of a degree.

association   A loosely organized group of stars, usually young, massive stars, that share a common origin. Also called OB association.

astronomical unit (AU)   The average orbital distance of the Earth from the Sun. This distance is 149,597,870.700 km (92,955,807.2730 mi), or more readily remembered as 150 million km (93 million mi). A parsec is equal to 206,265 AU—the number of arcseconds in a radian.

asymptotic giant branch (AGB)   Part of the Hertzsprung-Russell diagram where evolved stars are fusing helium and hydrogen in shells surrounding their dormant cores. Stars in this state are susceptible to pulsational instabilities and mass loss in the form of intense winds.

atom   The basic building block of ordinary gaseous, liquid, and solid matter. It consists of a central nucleus containing protons and neutrons, with a surrounding cloud of electrons. A neutral atom has equal numbers of oppositely charged protons and electrons.

atomic number   The number of protons and corresponding positive charge in the nucleus of an atom. Each element is specified by its atomic number.

atomic weight   The combined number of protons and neutrons in the nucleus of an atom. Different isotopes of the same element are differentiated by their atomic weight.

Balmer lines   The spectral lines of atomic hydrogen that appear in the visible part of the spectrum, either in absorption or in emission. They are produced by electronic transitions up from or down to the first energy level above the ground state ($n = 2$).

barred spiral   A galaxy with a distinctly linear central figure in its overall structure.

baryons   A class of heavy subatomic particles among the hadron family of elementary particles. Baryons include protons and neutrons, and are thought to consist of quark triplets.

Big Bang The hypothesized event some 14 billion years ago that exploded from a state of extremely high density and temperature to the universe we know today.

binary A double star whose two components revolve around each other's mutual center of mass.

black body An idealized object that absorbs all radiation that falls on it and re-emits it at a rate that maintains a state of thermal equilibrium. The spectrum of black-body radiation is completely specified by the body's temperature according to known laws.

black hole An object that is so massive and compact that neither light nor anything else can escape from its gravity.

Bremsstrahlung Radio wavelength continuum emission originating from ionized hydrogen, as free electrons accelerate past the hydrogen nuclei (protons).

bulge In a galaxy, the area around the nucleus of a spiral galaxy where stars are arranged in a less flattened volume than in the surrounding disk-like area. Galactic bulges typically contain old stars. The Milky Way's bulge also includes an admixture of young stars.

carbonaceous chondrite A type of stony meteorite that contains spheroidal granules rich in organic (carbon-rich) material, including some amino acids.

Cepheid variable A type of pulsating supergiant star with a period of variability of between about 1 and 100 days. The well-calibrated relation between a Cepheid variable's period and its absolute luminosity is used in establishing distances to nearby galaxies. The term was named for the prototype variable star, Delta Cephei.

Chandrasekhar limit The upper mass limit of a white dwarf, above which the stellar remnant will gravitationally collapse to a neutron star or explode. It was derived by Subrahmanyan Chandrasekhar in 1931 to be 1.4 solar masses.

cluster A physical group of stars that were born together or of galaxies that share a common origin.

Cold Dark Matter (CDM) Undetected material with significant mass and relatively low characteristic velocities. Includes ordinary (baryonic) matter in the form of planets and dead compact stellar remnants along with more exotic weakly interacting massive particles (WIMPs) of high mass.

collapsar   Hypothetical by-product of a massive rapidly rotating star that suddenly collapses to become a neutron star or black hole.

color index   A magnitude-based index of relative light output at different wavelengths from a celestial source. For example, the (B–V) color index is related to the ratio of observed fluxes (f) through B- and V-band filters according to $(B–V) = -2.5 \log (f[B]/f[V])$. In observations of stars, the (B–V) color and other color indices trace the temperature of the star's surface.

color-magnitude diagram (CMD)   A graphical plotting of colors and magnitudes of stars in a cluster or galaxy from which various characteristics of the stellar population can be inferred.

compact stellar object   The end-state of stellar evolution. Includes white dwarfs, neutron stars, and black holes.

continuous spectrum   The light emitted by an object at all wavelengths and without gaps caused by absorption at particular wavelengths. The rainbow is an example of a continuous spectrum.

corona   The hot outermost region of the Sun and other stars.

Cosmic Microwave Background (CMB)   The radiation left over from the epoch when the cooling universe changed from an ionized plasma to a neutral atomic state, approximately 380,000 years after the Big Bang.

cosmic rays   High-energy charged particles that originate beyond the Earth and are detected when they penetrate the Earth's atmosphere or when they interact with other components of the Milky Way.

cosmology   The study of the large-scale universe in terms of its origin, structure, and dynamics.

dark energy   An all-pervasive energy in the universe with no relation to any known radiating or absorbing objects. It could be driving the accelerating expansion of the universe that has been measured recently.

dark matter   Ponderous material (currently) of an unknown kind that is manifested in the gravitation necessary to bind the motions of stars in galaxies, of galaxies in clusters of galaxies, and of light rays traversing these massive objects.

dark nebula   A cloud of gas and dust that obscures the stars or galaxies behind it.

degeneracy   An unusual state of matter where the Pauli exclusion principle applies, so that the pressure depends on density alone, with no sensitivity to temperature. The electrons in a white dwarf star are degenerate, exerting sufficient pressure to support the star against gravitational collapse.

density wave   A periodic dynamical disturbance in the disk of a spiral galaxy in the form of two or more spiral wavefronts of enhanced density and gravity that can self-propagate around the disk and so generate the grand-design spiral structure that is observed in some spiral galaxies.

deuterium   A heavy isotope of hydrogen, H^2 (also known as D), where the nucleus contains one proton and one neutron.

differential rotation   The property of a rotating body whose rotation period is not the same at all radii, as opposed to solid-body rotation.

diffuse nebula   A cloud of gas and/or dust that is illuminated by starlight.

disk   In galaxies, the flat component of stars, gas, and dust in which spiral arms and associated star-forming activity are often present.

Doppler shift   The change in detected wavelength of light or sound caused by the motion of the source toward or away from the observer. For speeds significantly less than the speed of light, the Doppler shift $\Delta\lambda$ is quantified as $\Delta\lambda / \lambda = v/c$, where λ is the emitted wavelength, $\Delta\lambda$ is its shift, v is the velocity of approach or recession, and c is the velocity of light or sound.

dynamics   The gravitational interactions within a body or between bodies in motion.

electromagnetic radiation   Light in its many forms. This radiation involves co-oscillating electric and magnetic fields in the form of electromagnetic waves, whose energy propagates at the speed of light. The electromagnetic spectrum ranges from short wavelengths and high frequencies (gamma rays) to long wavelengths and low frequencies (radio waves).

electron   A negatively charged particle belonging to the lepton family of elementary particles, with a mass that is 1/1,836 that of a proton. Electrons either are bound to atoms or are unbound components of a plasma.

electron volt   The amount of energy acquired by an electron when accelerated through a voltage difference of one volt.

elliptical galaxy A type of galaxy that has a regular elliptical appearance and contains mostly very old stars yellow-red in color.

emission line The bright emission of a particular color (wavelength) in the spectrum of an incandescent or fluorescent gas.

emission nebula A gaseous cloud that emits the spectral emission lines of its constituent gases.

event horizon The critical surface surrounding a black hole where the velocity of escape equals the velocity of light.

exoplanet Object of planetary mass (1/1000 to 1000 times Earth's mass) residing beyond the Solar System—either in orbit around one or more stars, or traveling alone within its host galaxy. The range of masses is consistent with the body being massive enough for its gravity to impose a spherical shape but not so massive that it would be undergoing thermonuclear fusion in its core.

exponent The power to which a constant or variable is raised. Examples include 10^x, e^x, and x^n where x is the independent variable. Large exponents produce strong changes in the dependent variable.

extinction The reduction of the received intensity of light from a source caused by the absorption of that light by some intervening material, such as dust.

extremophiles Forms of life that thrive in conditions that would be regarded as hostile to most animals and plants. Early life on Earth may have been characterized by certain types of extremophiles.

extragalactic Beyond the realm of our local Galaxy.

fission The dissolution of atomic nuclei through the release of neutrons, protons, electrons, neutrinos, and associated gamma-ray photons. Isotopes of heavy elements such as uranium undergo fission at known rates.

flocculent spirals Spiral galaxies whose spiral structure is fragmentary and indistinct, reminiscent of sheep's wool.

flux A measure of received brightness in units of ergs per second per square centimeter at the detector, or equivalently Watts per square meter. The Sun's radiant flux on Earth is 1,380 Watts per square meter before absorption by the Earth's atmosphere. The brightest naked-eye stars have fluxes that are 10 billion times smaller.

forbidden lines Spectral emission lines that result from electronic transitions that are very improbable but nevertheless occur in the rarefied environment of interstellar space.

force   An interaction resulting in attraction or repulsion. The four known forces are the strong force between quarks in atomic nuclei, the weak force between quarks and leptons, the electromagnetic force between any charged particles, and the gravitational force between any masses.

frequency   The number of vibrations per unit of time, as in cycles/second (or Hertz [Hz]). A property of light waves.

fusion   The building up of atomic nuclei through the binding of light nuclei with other nuclei or nucleons. The Sun is thought to be powered by the thermonuclear fusion of hydrogen nuclei (protons) into helium nuclei (2 protons and 2 neutrons) and the associated release of gamma-ray photons and neutrinos.

galactic cluster   A small, relatively young (open) cluster of stars in our local Galaxy.

galaxy   A large collection of stars and other material components that is gravitationally bound, mostly autonomous with respect to other objects, and usually separated from its neighbors by millions of light-years. The Galaxy is our home galaxy, the Milky Way.

gamma rays   Very short wavelength, high-frequency, electromagnetic radiation. Gamma-ray photons are of very high energy.

general relativity   The theory of gravitation developed by Albert Einstein between 1906 and 1916, in which gravity is manifested by the curvature of space-time and the consequent trajectories of matter and light.

giant branch   The portion of the color-magnitude diagram where stars reside after their main-sequence lifetimes. The stars are typically brighter and often redder than they were on the main sequence.

gigahertz (GHz)   A unit of frequency equal to 10^9 Hertz (cycles/second).

giga-years (Gyr)   A unit of time equal to 10^9 years.

globular cluster   A usually massive, compact group of stars that is gravitationally stable. The globular clusters of the Milky Way are very old.

grand-design spiral   A spiral galaxy featuring prominent and continuous spiral arms in its disk.

gravitational instability   The property of an assemblage that determines its susceptibility to collapse under its own self-gravity.

gravitational lens   A massive object whose gravitational field bends the light from a background source.

ground state   The lowest allowed energy level in an atom ($n = 1$).

HI   Neutral atomic hydrogen gas.

HII   Ionized hydrogen gas.

HII region   A gas cloud that contains mostly hydrogen that is ionized but also glows in the light of ionized oxygen, sulfur, nitrogen, and other atomic species.

hadrons   A class of subatomic particles that are thought to be made of quarks and to be subject to strong interactions involving the exchange of gluons. Hadrons include baryons such as protons and neutrons, various kinds of mesons, and their antimatter counterparts.

halo   An outer envelope or constituent, especially of a galaxy. The Milky Way's halo contains mostly old metal-poor stars and ancient globular star clusters. The galactic halo is also thought to contain most of the dark matter in the Milky Way.

Herbig-Haro (HH) Objects   Dense ionized nebulae associated with star-forming regions. Thought to be shocked projectiles from proto-stellar and pre-main-sequence stars.

Hertz   A unit of wave frequency equal to one cycle per second.

horizontal branch   A portion of a color-magnitude diagram where very old stars are found at absolute visual magnitudes of about $M_V = 0.7$, after having been on the red-giant branch. These stars are thought to be fusing helium in their thermonuclear cores.

Hot Dark Matter (HDM)   Particles of low mass and relatively high characteristic velocities that interact weakly with ordinary matter and electromagnetic radiation. Neutrinos, if determined to have a mass, would be examples of this genre.

hypernova   An anomalously bright supernova, resulting from violent implosions of compact stellar remnants to black holes or collisions between these compact objects.

Index Catalogue (IC)   A supplement to the *New General Catalogue of Nebulae and Clusters of Stars* (NGC).

inclination   The angle between the plane of an object, such as that of a proto-planetary disk around a star and the imaginary plane of the sky. A face-on configuration has a zero inclination; an edge-on configuration has an inclination of 90 degrees.

infrared   Light of wavelength just longer than that of visible radiation.

The infrared band spans wavelengths of 1 micron (10^{-6} meters) to several hundred microns.

intensity   A measure of received brightness (flux) per unit angular area on the sky (as in Watts per square meter per square arcsecond), or, equivalently, luminosity per unit area at the source (as in solar luminosities per square parsec in a galaxy).

interstellar medium (ISM)   The gas and dust that exists between the stars.

ion   An atom that has lost or gained one or more electrons and is thus electrically charged.

ionization   The process of subtracting electrons from or adding electrons to previously neutral atoms, thereby changing a gas of neutral atoms to one made up of charged ions.

irregular galaxy   A type of galaxy with little or no symmetry or spiral structure.

isochrones   Lines of equal age in a color-magnitude diagram.

isotope   A variation of an atom, or element, of fixed atomic number that depends on the total number of protons and neutrons (the atomic weight). For example, the isotopes of uranium-238 and uranium-235 differ by their respective numbers of neutrons, the number of protons being 92.

isotropic   Having properties that do not depend on direction.

Jeans length   The characteristic size of a perturbation that is just large enough to be gravitationally unstable.

Jeans mass   The mass of a perturbation of a given size that is just large enough to be gravitationally unstable.

Kepler's laws   Three laws of planetary motion discovered by Johannes Kepler. (1) Each planet orbits the Sun along an elliptical trajectory. (2) It speeds up when near the Sun and slows down when farther from the Sun, such that it sweeps out equal areas in equal times. (3) The average velocity of each planet decreases as the square root of the ellipse's semi-major axis. This is noted as the square of the planet's orbital period (in years) equaling the cube of the semi-major axis of the planet's orbit (in AU).

kiloelectron volts (KeV)   One thousand electron volts, a measure of energy.

kiloparsec (kpc)   One thousand parsecs, amounting to 3,260 light-years.

kinematics   The motions of objects, as in the case of the stars in a galaxy.

leptons   A class of light subatomic particles that includes electrons, neutrinos, and their antimatter counterparts. Leptons are subject to weak, electromagnetic, and gravitational interactions but not to strong interactions.

light curve   A graphical plot of the changing brightness of a source over time.

light-day   The distance light travels in an Earth day (2.6×10^{10} kilometers).

light-year   The distance that light travels in one Earth year (9.46×10^{12} kilometers).

Local Bubble   A relative void in the interstellar medium local to the Solar System and extending for several hundred light-years. Very low-density million-degree gas fills this void. Both the void and the hot gas are thought to be the result of one or more supernova explosions during the last few million years.

Local Group   The family of about 40 galaxies to which the Milky Way Galaxy belongs.

Local Standard of Rest   The reference frame centered about the Solar System in circular orbit around the galactic center. The velocity of this reference frame is determined from the average velocity of stars and gas in the solar neighborhood, after removal of the Sun's peculiar motion.

logarithm   The power of 10 that generates a particular number. For example, the number 1,000 can be stated as 10^3, its logarithm being 3. The fraction 1/1,000 can be stated as 10^{-3}, its logarithm being -3.

lookback time   The time between the present observation of a distant source and the time when the light from the source was emitted. The lookback time for the Sun is about 8 minutes.

luminosity   The intrinsic power of a radiating source, typically measured in units of ergs per second, Watts, or solar luminosities. Radiant luminosities refer to the output of photons from a source. Mechanical luminosities refer to the amounts and speeds of matter flowing from a source.

luminosity class   An estimate of the luminosity of a star based on certain observed criteria that divide stars into recognizable classes. Main-sequence (Dwarf) stars are of luminosity class V, giant stars are of luminosity class III, and supergiant stars are of luminosity class I.

luminosity function   A curve or table that tells how many stars or galaxies there are at each absolute luminosity.

Magellanic Clouds   The Large and Small Magellanic Clouds are irregular galaxies that orbit the Milky Way Galaxy. Observable from the southern hemisphere, they are the largest satellite galaxies of the Milky Way.

magnetar   A type of neutron star with extremely intense magnetic fields. Reconfigurations of these magnetic fields may explain X-ray and gamma-ray outbursts that have been observed.

magnitude   A measure of the brightness of an astronomical object that uses a logarithmic scale. An increase of 1 magnitude corresponds to a diminution in brightness by a factor of 2.5, and an increase of 5 magnitudes corresponds to a dimming by a factor of $(2.5)^5$, or 100.

Main Sequence   The diagonal band in a color-magnitude or temperature-luminosity diagram where most stars are found. It represents the locus of temperature and luminosity that characterizes stars of different mass during the most stable and long-lasting part of their lives.

mass   The measure of the amount of matter in an object as determined by its gravitational field or by its response to a force.

Massive Compact Halo Objects (MACHOs)   Conjectured to be the dark matter that is concentrated in galaxies and galaxy clusters. Includes isolated planets, dead compact stellar remnants, and more exotic massive subatomic particles that do not absorb or emit significant electromagnetic radiation.

mega electron volts (MeV)   A unit of energy equal to 1 million electron volts (eV).

megaparsec (Mpc)   A unit of distance equal to 1 million parsecs, amounting to 3,260,000 light-years.

mega-year (Myr)   A unit of time equal to 1 million (10^6) years.

metal   In the parlance of astronomers, any element heavier than helium. Carbon, oxygen, and iron provide observable indices of a region's metallicity.

metal-poor   Lacking in heavy elements compared to the Sun.

metal-rich   Similar to the Sun in terms of heavy element abundance.

micrometer   A measure of length equal to one-millionth of a meter.

micron   A measure of length equal to one-millionth of a meter.

microwave   Short-wavelength radio waves with wavelengths between about 1 millimeter to 300 millimeters (30 centimeters).

Milky Way Galaxy   A self-gravitating assemblage of stars, planets, gas, dust, and suspected dark matter (still of unknown provenance) belonging to the Local Group of galaxies. By mass, the Milky Way is estimated to be 83% dark matter and 16% stars, with the remaining 1% consisting of gas (0.99%) and dust (0.01%) for a grand total of about a trillion solar masses. These constituents are arranged in varying proportions within the bulge/bar, disk, and spheroid (halo) components. Also known as the Galaxy, the Milky Way is classified as a SBbc barred-spiral galaxy.

molecule   An electrically bound assemblage of atoms.

molecular cloud   A cool, dense cloud of gas and dust in which molecules are a conspicuous component. Most molecular clouds are dominated by diatomic hydrogen (H_2) but glow most brightly in emission lines of carbon monoxide (CO).

nebula   A diffuse cloud, usually of gas or dust or both. Historically, nebulae included any extended fuzzy-looking object, including what we today recognize as galaxies. Today, the term nebula is restricted to cold, dark molecular clouds, cool, diffuse "cirrus" clouds, and hot, ionized HII regions, Herbig-Haro objects, planetary nebulae, and supernova remnants.

neutrino   A very small subatomic particle belonging to the lepton family, with no charge, little or no mass, and poor interactivity involving the weak force.

neutron   A subatomic particle belonging to the hadron family with no charge and with a mass slightly more than that of a proton. Outside of the atomic nucleus, the neutron disintegrates into a proton, electron, and neutrino in about 11 minutes.

neutron star   A very dense and compact stellar remnant following the collapse of a massive star's core, or perhaps of a white dwarf remnant that has been accreting mass from a nearby companion star. The collapsed core is comprised primarily of neutrons and has a density similar to that of atomic nuclei. The lower mass limit of a neutron star is 1.4 solar masses, and the upper limit is about 3 solar masses—above which the object would collapse to a black hole.

New General Catalogue of Nebulae and Clusters of Stars (NGC)   A catalog completed near the end of the nineteenth century—begun by William Herschel and further developed by his son John Herschel.

nova   A star that suddenly increases in brightness owing to the exchange of gases from a more massive star to a small, dense, hot companion star, resulting in a runaway thermonuclear reaction on the dense star's surface. White dwarfs are thought to be the hosts of most nova outbursts.

nucleic acid   A macromolecule found in all living cells, where they function in encoding, transmitting, and expressing genetic information. Ribonucleic acid (RNA) is composed of a twisted sugar-phosphate backbone studded with nitrogenous bases. Deoxyribonucleic acid (DNA) is composed of two sugar-phosphate backbones twisted into a helix with nitrogenous base pairs connecting them like rungs on a twisted ladder. These molecular components are known as nucleotides.

nucleosynthesis   The process in which larger atoms are built up from smaller ones through thermonuclear fusion.

nucleus   The central regions of an object, such as a galaxy.

OB association   A group of very young and hot stars that formed together but that are not necessarily gravitationally bound.

open cluster   A group of stars, usually smaller, less condensed, and younger than a globular cluster. Often called a galactic cluster.

parallax   The apparent annual back-and-forth motion of a nearby star with respect to background stars in the sky, resulting from the Earth's orbital motion around the Sun.

parsec   A unit of distance equal to that of a star that exhibits a heliocentric parallax over a quarter year of 1 second of arc (1 arcsecond). One parsec equals 3.26 light-years, or 3.086×10^{13} kilometers. It is the abbreviation of "parallax arcsecond."

Pauli exclusion principle   A property of subatomic matter that excludes two particles of the same type from occupying the exact same quantum state. It is obeyed by fermions (baryons and leptons), but not by bosons (photons and mesons). It explains how white dwarfs and neutron stars can keep themselves propped up against their strong gravitational fields.

peculiar velocity   The velocity of an object that is different from the average velocity of the objects in its surroundings. The Sun's peculiar velocity must be considered when determining the local standard of rest.

period-luminosity relation   The observed relationship between the periods of pulsation and the absolute luminosities of Cepheid variable stars.

perturbation   A deviation from the normal, as in the orbital motion of an object or objects or in the distribution of material in a medium.

photo-evaporation   The process whereby ultraviolet photons from a hot star or active galactic nucleus ablate material off the surface of an exposed nebular region.

photo-ionization   The process involving the ionization of a gas by energetic photons.

photometry   The measurement of the brightness of an object.

photon   A quantum particle of electromagnetic radiation (light) having no mass or charge. The energy of a photon is proportional to the frequency of the corresponding light wave, and inversely proportional to the wavelength. The photon conveys the electromagnetic force.

photosphere   The layer of a star, from which photons of light can freely escape. It is the surface we see.

planetary nebula   An ejected shell from a highly evolved star that is irradiated by the exposed hot stellar core and made to fluoresce.

plasma   A gas consisting of ionized atoms and free electrons. The charges of these ions and electrons make the gas especially sensitive to magnetic fields and electromagnetic radiation (light).

plunging orbit   An elliptical orbit with a very large eccentricity, or an unbound hyperbolic orbit, that carries the object in an almost radial path to and from the center. Globular clusters and stars belonging to the galactic halo often follow plunging orbits.

polycyclic aromatic hydrocarbons (PAHs)   Organic molecules that are based on the benzene ring molecule, C_6H_6, and are detected in circumstellar and interstellar environments—typically at infrared wavelengths.

potential energy   The energy inherent in a configuration that has yet to be actualized.

pre-main-sequence (PMS) stars   The transitional stage between protostars and main-sequence stars characterized by core deuterium burning and additional luminosity from gravitational contraction of the object.

proper motion   The apparent motion of a star in the sky with respect to background stars or galaxies, measured in arcseconds per year. Nearer stars typically have larger proper motions.

protogalaxies   Objects in the universe that evolve to become galaxies.

proton   A positively charged subatomic particle that belongs to the hadron family of elementary particles and exists in atomic nuclei.

protoplanetary disks (proplyds) The nebular disks around protostars and pre-main-sequence stars that precede the formation of planets.

protoplanets Objects in galaxies typically associated with protostars and pre-main-sequence stars that evolve to become planets.

protostar A star that has not yet reached its equilibrium state on the main sequence. Gravitational accretion of surrounding matter powers its radiation, which is typically detectable at submillimeter and infrared wavelengths.

pulsar A rotating magnetized neutron star that emits radio and optical light pulses in a manner similar to that of a lighthouse.

quantum Pertaining to subatomic particles, where uncertainties in position, momentum, time, and energy are important, and where observed behavior is explained in terms of wave physics and probabilistic theories. Also pertaining to discrete allowed configurations of specified charge, mass, angular momentum, and energy—as in quanta of light (photons) and the quantized states of atoms.

quark A subatomic particle of fractional charge whose various combinations make up the hadron family of particles, including protons, neutrons, and mesons. Quarks are found only in these combinations— never in isolation. They are especially subject to the strong force, but also respond to the weak, electromagnetic, and gravitational forces.

quasar The brilliant, small source of light found in certain distant galaxies and resulting from the presence of a supermassive black hole in the Galaxy's nucleus that is accreting its immediate surroundings and radiating away the gravitational energy that is released during the infall.

radial velocity The measured velocity along the line of sight—typically ascertained by virtue of the Doppler shifting of emission or absorption lines in the spectrum of the object.

reddening The selective absorption of shorter-wavelength light by the interstellar medium, resulting in the selective transmission of longer-wavelength (redder) light. The setting Sun is reddened by the Earth's atmosphere in a similar way, such that blue and green light is absorbed or scattered into the sky, leaving orange and red light to shine through.

red giant A star that has expanded and cooled following its main-sequence phase. It is thought to be fusing hydrogen in a shell that surrounds its dormant core.

redshift The Doppler shifting of observed light to longer (redder) wavelengths, caused by the recession of the source.

relativity   The formulations of natural law that incorporate the constancy of the velocity of light in any reference frame (special relativity) and the relation between gravity and the curvature of space-time (general relativity).

relaxation   The dynamical smoothing and rearranging of stars over cosmic time in a system of mutual gravitational interactions.

resolution   The level of detail that can be discerned in an image, spectrum, or light curve.

RR Lyrae variable star   A short-period pulsating giant star, commonly found in old globular clusters, whose near-constant average brightness is used as a standard candle to gauge the distances of these clusters.

Schwarzschild radius   The radius demarcating the event horizon of a non-rotating spherical black hole, inside of which light or any other form of matter-energy is unable to escape.

self-gravitation   The property of a body that is dominated by its own gravitational field, exclusive of outside influences.

shock ionization   The process in which the atoms in a gas are collisionally ionized by the passage of a shock front that rapidly compresses and energizes the gas.

solar neighborhood   The immediate area around the Sun, spanning approximately 100 light-years in radius, and including the nearest few thousand stars.

space-time   An inclusive term that relates the three dimensions of space and one dimension of time in the presence of gravity and other energy fields. The general theory of relativity explicitly uses space-time in its formulations.

spectral energy distribution (SED)   The relationship between the amount of energy emitted from a luminous source and the wavelength of the emitted light.

spectral line indices   Measures of the strengths of spectral lines of certain elements in the spectrum of a star or galaxy. HII regions show strong line emission from singly and doubly ionized elements, while young supernova remnants and active galactic nuclei show strong emission from elements at higher states of ionization.

spectral type   A classification of the spectrum of a star according to which lines are prominent. A star's spectral type is related to the surface temperature and gravity of the star.

spectroscope   An instrument for recording the spectra of objects.

spectrum   The light of a source, spread out into its different wavelengths.

starburst   A galaxy or portion thereof that is experiencing an energetic episode of enhanced star formation involving large numbers of hot massive stars and associated supernova activity.

stochastic   Governed by randomly occurring processes.

subdwarf   A type of main-sequence star that is metal poor. With fewer metal absorption lines in the blue part of its spectrum, it appears bluer than most main-sequence stars of solar metallicity. Therefore it occupies a part of the Hertzsprung-Russell diagram that is shifted blueward of the nominal Main Sequence.

superbubble   A nebular void excavated by UV photons and winds from hot stars along with supernova blast waves from recently expired massive stars.

supercluster   An unusually large or massive cluster of either stars or galaxies.

supergiant   An evolved, massive star that is brighter than the more common giant stars.

supernova   The explosive destruction of a massive star, following the implosion of its iron core to a neutron star or black hole (Type II). Alternatively, the explosive destruction of a white dwarf stellar remnant whose mass exceeds the Chandrasekhar limit of 1.4 solar masses (Type Ia). A supernova is observed as an extreme brightening of the original stellar source.

supernova remnant (SNR)   The gaseous cloud of material ejected by the explosion of a supernova and/or of interstellar material that has been plowed up by the supernova's blast wave.

synchrotron radiation   Light emitted by charged particles moving at velocities close to that of light in the presence of a magnetic field. The radiation has a unique nonthermal spectrum and is highly polarized.

thermal   Pertaining to an ensemble of objects with a common temperature. Thermal radiation is that emitted by such an ensemble. Its spectrum is similar to that of an idealized black body.

thick disk   A component of the Galaxy's disk that is several times thicker than the thin disk but ten times less populated by stars. The origin of the thick disk remains uncertain.

thin disk   The main component of the Galaxy's disk, containing most of the disk's stars, gas, and dust. It extends roughly 1,000 light-years above and below the midplane of the Milky Way.

tidal interactions   The effects of gravitational tides of one body on another where the tides involve differential gravitational forces across each of the interacting bodies.

velocity dispersion   The range in velocities seen, for example, in the stars occupying the center of a galaxy, or among stars in the disk of a galaxy.

velocity of escape   The velocity that an object must have away from another body in order to become clear of the gravitational attraction of the other body.

V_{LSR} The (radial) velocity of an object with respect to the Local Standard of Rest.

water hole   A part of the electromagnetic spectrum, where emission from the Milky Way and larger cosmos is relatively faint, thus allowing for easier detection of artificial signals from extraterrestrial sources. This radio domain includes the spectral line of atomic hydrogen (H) at a wavelength of 21 centimeters and spectral lines from the hydroxyl molecule (OH) at wavelengths near 18 centimeters. These two emitting species make up the molecule of water (H_2O), hence the term.

Weakly Interacting Massive Particles (WIMPs)   Hypothetical particles that show negligible interactions with ordinary matter and electromagnetic radiation. WIMPs may constitute a significant fraction of the dark matter in the universe.

white dwarf star   The collapsed remnant core of an evolved star of intermediate mass that has exhausted its nuclear fuel. The core is made up of helium and carbon nuclei and is held up by the degenerate pressure exerted by charged electrons. The upper mass limit of a white dwarf is 1.4 solar masses, above which the core collapses to a neutron star or black hole, or explodes.

Wolf-Rayet star   An extremely hot, massive star that is ejecting gas shells at high velocity.

X-rays   High-energy electromagnetic radiation with wavelengths of about 10 Angstroms to 100 Angstroms—shorter than those of ultraviolet light but longer than those of gamma rays.

NOTES

· · ● ● ● · ·

CHAPTER 2: HISTORIC PERCEPTIONS

1. After marrying the Russian astronomer Sergei Gaposhkin in 1934, Cecilia Payne became Cecilia Payne-Gaposhkin. Her subsequent legacy of research and writing all bear her married name. She was arguably one of the first prominent female scientists to deal with these sorts of authorship issues.

2. Shapley's interests ranged far and wide. He discovered the first dwarf spheroidal galaxies, authored several general-interest books on astronomy (including the first book on galaxies), built up a thriving astronomy department at Harvard, helped to found the National Science Foundation (NSF) and the United Nations Educational, Scientific and Cultural Organization (UNESCO), defended international cooperation among scientists despite harassment by the notorious House Committee on Un-American Activities, published extensively researched papers on ants, and fathered a remarkable brood of scientists, writers, and educators. Hubble stuck to his nebulae, developing new "standard candles" for determining distances out to the most remote sources recorded on his photographic plates. In so doing, he found a relation between the spectral redshifts of these objects and their distances. Moreover, he brilliantly interpreted this relation as the unique consequence of an expanding universe. Our conception of the cosmos has not been the same since.

CHAPTER 3: PANCHROMATIC VISTAS

1. The conclusion that the Milky Way harbors vast amounts of dark matter provides some insights to the nature of this elusive agent in the cosmos. If the particles making up the dark matter were neutrinos or other light particles that don't interact electromagnetically, they would be

traveling too fast to be bound by the Milky Way's overall gravitation. Relativistic neutrinos and other forms of Hot Dark Matter are simply too itinerant to make up the Milky Way's quota of dark matter. Weakly interacting massive particles (WIMPs) such as neutralinos (the supersymmetric counterparts of neutrinos in unified theories of matter) are sufficiently heavy and correspondingly slow enough to remain confined within the Galaxy's halo. This form of Cold Dark Matter is also consistent with current theories of galaxy formation (see chapter 11). Massive halo compact objects (MACHOs) include isolated planets, brown dwarf stars, faint old white dwarfs, neutron stars, and black holes. MACHOs provide an alternative form of Cold Dark Matter—one that we know exists to some degree in the galaxy. However, the total amounts are probably far from what is required to make up the roughly 300 billion solar masses that fill the galactic halo. We are thus left with some type of WIMP or other exotic "dark" particle. Detection of a single WIMP or other subatomic candidate has yet to be confirmed, however, despite several very expensive experiments directed toward this important goal.

2. The human eye is able to detect light over a 0.4–0.7 micron range of wavelengths. Modern electronic detectors can do much better, however, increasing the nominal "optical" range to 0.3–1.0 microns.

CHAPTER 4: NEIGHBORS OF THE SUN

1. A more popular argument among contemporary scientists is that the vast and chemically complex galaxy we happen to live in was necessary for our emergence. Therefore, it is no surprise that we find ourselves in such a well-endowed system. This argument is a form of "anthropic reasoning" which contends that the universe is the way it is, because if it were significantly different we would not be here. Other universes may vary.

2. Both units of distance are Earth-centric, in that one is based on the Earth's orbital radius (the parsec), and the other is based on the Earth's orbital period (the light-year). Other planetary civilizations would likely have their own home-based units for defining interstellar and intergalactic distances.

3. With its origins in the naked-eye observations of Hipparchus circa 5 BCE, the magnitude system of brightnesses compresses the large dynamic range of illumination by considering ratios. The modern magnitude system, first introduced by Norman Pogson in 1856, specifies that a differ-

ence of one magnitude corresponds to a ratio of 2.5 in brightness. A difference of 5 magnitudes then translates to a factor of $(2.5)^5$ or 100 in brightness, and a difference of 10 magnitudes nets a full $(2.5)^{10}$ or 10,000-fold brightness ratio. The formulation that relates magnitudes and brightnesses (or fluxes [f]) is then

$$m_2 - m_1 = -2.5 \log (f_2 / f_1),$$

where "log" refers to taking the base-10 logarithm, the fluxes are in physical units of Watts/m^2 or equivalent representation, and the minus sign ensures that the greater the magnitude, the fainter the source. The naked-eye sky of Hipparchus spanned a range from the brightest zeroth magnitude stars to the faintest appearing sixth magnitude stars. The most powerful contemporary telescopes can sense thirtieth magnitude sources, a trillion times fainter than the brightest stars in the night sky.

4. You do not have to struggle to see the stellar hosts of exoplanets, as several dozen of them are visible in the naked-eye sky. Brightly shining at first magnitude, Pollux in Gemini is known to have a "hot Jupiter" orbiting uncomfortably close to it. Algieba, the second magnitude star at the base of Leo's neck, has at least one planet, possibly two. And 55 Cancri at the sixth-magnitude naked-eye limit, hosts a system of five planets, one of which orbits within the star's habitable zone.

5. Spectral resolution is defined as the minimum observable change in wavelength compared to the wavelength itself. It is commonly expressed as a ratio, with the nominal wavelength appearing in the numerator, i.e.

Resolution = (nominal wavelength) / (change in wavelength), or

$$R = \lambda_o/\Delta\lambda \, .$$

By virtue of the Doppler effect, this resolution also corresponds to the minimal velocity shift that can be detected, according to

$$R = c/\Delta v,$$

where c is the speed of light. In principle, a velocity resolution of 3 m/s would necessitate a spectroscopic resolution of 100 million, or equivalently 1/18,000 of an Angstrom! This level of precision is very difficult to achieve. Recent techniques make use of multiple spectral lines from the target star that are cross-correlated with a highly accurate template spectrum. In this way, a spectrometer with a resolution of "only" a few hundred thousand can yield velocity resolutions of 3 m/s or finer.

6. Such impacts by massive clouds are certainly possible. For example, the giant cloud of atomic hydrogen known as Smith's Cloud appears to be on a collision course with the Milky Way (see figure 9.12). First discovered in 1963 by the American astronomer Gail Smith, the cloud has been recently mapped in its entirety. It measures 11,000 × 2,500 light-years in extent, contains about a million Suns worth of atomic hydrogen, and is speeding toward the Milky Way at 240 km/s on a trajectory that is inclined 45 degrees with respect to the galactic plane. At an estimated distance of 40,000 light-years, Smith's Cloud will plow into the Milky Way within the next 20–40 million years. The ensuing repercussions should be as impressive as those evident in Gould's Belt today.

CHAPTER 5: BEACONS FROM AFAR

1. The trigonometric relationship between transverse velocity (v_{trans}) and radial velocity (v_{rad}) of a star in a "moving cluster" is

$$v_{trans} = v_{rad} \times \tan(\text{angle}),$$

where the transverse velocity (v_{trans}) is

$$v_{trans} = \mu \times d,$$

with the proper motion μ expressed in radians/second, the distance d in kilometers, the velocity v_{rad} in kilometers/second, and the angle referring to that between the star and the convergent point of the moving cluster. After conversion to handier units, one obtains the following solution for distance:

$$d = v_{rad} \times \tan(\text{angle}) / (4.74 \times \mu''),$$

where the distance d is in parsecs, the radial velocity v_{rad} is in kilometers/second, and the proper motion μ'' is in arcseconds/year. The average of stellar distances as obtained this way provides a reliable measurement of the cluster's distance.

2. A "black body" is an idealized object that absorbs all radiation that falls on it and re-emits it at a rate that ensures a state of perfect thermal equilibrium. The spectrum of black-body radiation is completely specified by the body's temperature according to known laws (see figure 3.6).

Moreover, the total radiative output depends solely on the body's size and temperature. For a spherical black body approximating a star, the luminosity L depends on the body's spherical surface area ($4 \pi R^2$, where R is the radius of the sphere) and on the 4th power of its surface temperature (i.e., $L = 4 \pi R^2 \sigma T^4$, where σ is the Stefan-Boltzmann constant for black-body radiation [$\sigma = 5.67 \times 10^{-8}$ Watts/m^2-K^4]). For example, a yellow-giant star with the same surface temperature as that of the Sun but with a 100 times larger radius would be 100^2 or 10,000 times more luminous according to this black-body formulation. If another star was the same size as the Sun but ten times hotter, it would be 10^4 or once again 10,000 times more luminous.

3. The distance modulus ($m - M$) that compares the apparent magnitude m to the absolute magnitude M is related to the actual distance according to

$$(m - M) = 5 \times \log(d/10),$$

where the distance d is in parsecs, and "log" represents the base-10 logarithm. Solving for the distance yields

$$d = 10^{(1 + 0.2 [m-M])}.$$

These magnitude-based relations can be directly derived from the more physical "inverse-square" law that relates measured fluxes to luminosities according to

$$f = L / (4 \pi d^2),$$

where the flux f is in units of Watts/meter^2, the luminosity L is in Watts, and the distance d is in meters.

CHAPTER 6: STAR BIRTH

1. This lovely simile comes from Timothy Ferris's evocative description of HII regions in his book *Galaxies*.

CHAPTER 7: LIVES OF THE STARS

1. Far more comprehensive stellar biographies can be found in the books and websites on stars by James Kaler (see the Bibliography in this book and http://stars.astro.illinois.edu/sow/sowlist.html).

2. Stars with masses between 0.01 and 0.08 solar masses are too insubstantial to host hydrogen fusion in their cores. The cores are simply not hot and dense enough. However, central conditions are sufficiently energetic to permit the fusion of deuterium (a proton and a neutron) into helium. Such stars are known as *brown dwarfs*. Because of their very low luminosities, they have been difficult to find. The first brown dwarfs were confirmed in 1995, and only recently they have been detected in any abundance. Extrapolating from these findings, astronomers suspect that the net number of brown dwarfs in the Milky Way may comprise a significant fraction of the total stellar population by number but not by mass.

3. This formulation makes explicit the equivalence between pressure and energy density, as (n) is the particle number density and $(k\,T)$ is none other than the average thermal energy per particle.

CHAPTER 8: STELLAR AFTERLIVES

1. White dwarfs are sometimes called "diamond stars" because they are made of lots of carbon in a crystalline state akin to that of diamonds. However, the densities of these two substances differ by factors of more than 100,000. Moreover, the electrons in a white dwarf are not bound by individual nuclei—like they are in the carbon atoms that comprise a diamond crystal. Instead, they are free to roam throughout the dense matrix of carbon, oxygen, and helium nuclei—thereby making the white dwarf highly conductive and magnetic.

2. European scribes also noted a bright star appearing in the sky near the constellations of Taurus and Gemini, but at the earlier date of April 11, 1054—nearly coinciding with the death of Pope Leo IX on April 19. Because of uncertainties and errors in these various historical accounts, we may never know the exact date of the Crab supernova's first appearance.

CHAPTER 9: THE GALACTIC GARDEN

1. The *Local Standard of Rest*, or LSR, refers to the Sun's co-moving neighborhood of stars. It is defined as the reference frame, instantaneously centered on the Sun, which moves in a circular orbit about the galactic center at the speed appropriate to its position in the galaxy (about 220 km/s). The determination of a source's radial velocity with

respect to the LSR (v_{LSR}) involves spectroscopic measurements of the source, translation of the Doppler-shifted line emission to a line-of-sight velocity, and removal of the Sun's peculiar motion in that direction relative to the LSR.

CHAPTER 10: MONSTER IN THE CORE

1. The light from galaxies and quasars that are billions of light-years away has taken billions of years to travel to our telescopes on Earth. Therefore, we see these objects as they were billions of years ago. This time lag between emission and detection is known as the "lookback time." We can infer that these objects are at such fantastic distances by gauging how much their light waves have been stretched by the ever-expanding universe through which the waves have been propagating. A galaxy whose light waves have been stretched by factors of 2, 3, and 4 have corresponding spectral "redshifts" of 1, 2, and 3—where the redshift z is defined to be the shift in observed versus emitted wavelength ($\lambda - \lambda e$) divided by the emitted wavelength (λe), so that $z = (\lambda - \lambda_e)/\lambda_e$, and the degree of stretching due to universal expansion is $(R/R_e) = \lambda/\lambda_e = 1 + z$, where R represents the scale of the universe at the present epoch, and R_e represents the scale at some earlier epoch that corresponds to the emitting object. The corresponding lookback time in a uniformly expanding universe is

$$t - t_e = t\{1 - (1/[1 + z])\},$$

where t is the age of the current epoch, and t_e is the age of the universe at some earlier epoch as traced by the emitting object. For a universe that is 14 billion years old, objects at redshifts of 1, 2, and 3 would be observed at lookback times of 7, 9, and 10 Gyrs, respectively.

CHAPTER 12: LIFE IN THE MILKY WAY

1. In her review on "The Search for Extraterrestrial Intelligence" (*Annual Reviews of Astronomy and Astrophysics* 39: 511 [2001]), pioneering SETI scientist Jill Tarter cites the following definition of life: "Life is a self-sustained [chemical] system capable of undergoing Darwinian evolution."

2. A prior version of these ruminations over the origins of water on

Earth first appeared as an online white paper by the author as part of the *Year of Water* hosted by Sigma Xi, the Scientific Research Society (see http://www.sigmaxi.org/programs/issues/water.shtml).

3. This treatment of the Drake Equation uses the total number of stars $N(star)$ rather than the star formation rate $R(star)$ that Frank Drake originally used. I think that $N(star)$ is more pertinent to the question at hand—namely, the number of stars that are *currently* suitable to hosting habitable planets. In order to end up with the desired number of technologically communicative planets, I use a fraction for the last factor rather than the total lifetime of technologically communicative civilizations that Drake used. In estimating this fraction, the total lifetime of technologically communicative civilizations ends up being invoked. A prior version of this treatment first appeared as part of the website http://cosmos.phy .tufts.edu/cosmicfrontier that complements the book *Galaxies and the Cosmic Frontier* by William H. Waller and Paul W. Hodge.

SELECTED READINGS

•• • • ••

HISTORY

Belkora, Leila. *Minding the Heavens: The Story of Our Discovery of the Milky Way*, Philadelphia, PA: Institute of Physics Publishing, 2003. Provides case studies of galactic astronomy in the seventeenth through early twentieth century with biographies of Thomas Wright, William Herschel, Wilhelm Struve, William Huggins, Jacobus Kapteyn, Harlow Shapley, and Edwin Hubble.

Berendzen, Richard, Richard, Hart, and Daniel Seeley. *Man Discovers the Galaxies,* New York: Science History Publications, 1976. A well-illustrated history of the astronomers and the advances that they made in galactic and extragalactic astronomy—from William Herschel to Edwin Hubble.

Croswell, Ken. *The Alchemy of the Heavens: Searching for Meaning in the Milky Way*, New York: Anchor Books, 1995. Focuses on the many advances that occurred during the twentieth century, with lively accounts of the astronomers who helped build our current understanding of the Milky Way.

Galilei, Galileo. *Sidereus Nuncius*, Venice, 1610. Trans. A. van Helden as *The Starry Messenger*, Chicago: University of Chicago Press, 1989. Galileo describes his pioneering telescopic observations of the Moon, Venus, Jupiter, the Sun, star clusters, and the Milky Way.

Herschel, William. *On the Construction of the Heavens* (1785), in *The Scientific Papers of Sir William Herschel*, London: The Royal Society and the Royal Astronomical Society, 1912. Includes the observations by William and Caroline Herschel of nebulae, their star-gauging efforts, and their resulting map of the Milky Way as viewed from outside.

Krupp, Edwin C. "Negotiating the Highwire of Heaven: The Milky Way and the Itinerary of the Soul," in *Vistas in Astronomy* 39 (1995): 405–430. An examination of pre-Copernican and contemporary indigenous spirituality associated with the Milky Way.

Krupp, Edwin C. "Spilled Milk," in *Griffith Observer* 57, no. 12 (1993): 2–18. Interpretations of the Milky Way as perceived by various cultures. Includes many illustrations with extensive captions.

Lesko, Barbara S. *The Great Goddesses of Egypt*, Norman: University of Oklahoma Press, 1999. Includes detailed descriptions of the sky goddess Nut within the pantheonic context of pre-dynastic and dynastic Egypt.

Struve, Otto, and Velta Zebergs. *Astronomy of the 20th Century*, New York: Macmillan, 1962. A masterful compendium of the astronomical research and advances that occurred in the first half of the twentieth century. Written by an expert astronomer of distinguished lineage and his able research assistant.

Whitney, Charles A. *The Discovery of our Galaxy*, New York: Alfred A. Knopf, 1971. A thorough and rather philosophical account of Milky Way astronomy, beginning with Nicholas Copernicus's heliocentric model of the Solar System and Giordano Bruno's heretical speculations on an infinite universe of worlds beyond the Solar System. Profiles of Galileo Galilei, Thomas Wright, the Herschels, and subsequent astronomers are enriched with revealing quotations.

GENERAL INTEREST

Bally, John and Bo Reipurth. *The Birth of Stars and Planets,* Cambridge, UK: Cambridge University Press, 2006. A richly illustrated account of the early stages of stellar evolution—from cold, dusty molecular clouds in the Milky Way to protostars, proto-planetary systems, and the interstellar consequences of short-lived massive stars.

Bok, Bart J. and Priscilla F. Bok. *The Milky Way*, fifth edition, Cambridge, MA: Harvard University Press, 1981. The "classic" nontechnical book on the Milky Way and its contents, written by one of the most beloved husband-wife teams in astronomy. Formed part of the Harvard Book Series on Astronomy.

Burnham, R., Jr. *Burnham's Celestial Handbook,* Mineola, NY: Dover Publications, 1978. This three-volume masterpiece is organized according to constellation. Each constellation heading includes descriptions of resident stars, clusters, nebulae, and galaxies along with pertinent mythologies, poems, and astronomical histories. Essential reading for all amateur and armchair astronomers.

Dartnell, L. *Life in the Universe, A Beginner's Guide,* Oxford, UK: One-World Publications, 2007. A contextual approach to astrobiology starting with the marvelous "workings of life" on Earth and moving out to explore Earth's cosmic origins, the evolution of life on Earth, and the conditions for life elsewhere in the Solar System.

Ferris, Timothy. *Galaxies*, New York: Stewart, Tabori & Chang Publishers, 1980. A well-illustrated and eloquently written coffee table book on galaxies, with the first eighty pages dedicated to the Milky Way.

Henbest, Nigel, and Heather Cooper. *The Guide to the Galaxy*, Cambridge, UK: Cambridge University Press, 1994. A graphic introduction to the Milky Way as a spiral galaxy of stars, gas, and dust. Includes artistic renderings of the local bubble, neighboring spiral arms, and nucleus, based on best estimates of the distances to various sources in these features.

Kaler, James B. *Stars*, New York: Scientific American Library, 1998. A complete description of stars, their observed features, and life histories written by a professional aficionado with color illustrations throughout. Dr. Kaler also tends a website on stars which is available at http://stars.astro.illinois.edu/sow/sowlist.html.

Kaler, James B. *Cosmic Clouds: Birth, Death, and Recycling in the Galaxy*, New York: Scientific American Library, 1997. A colorful exposition of the fertile "stuff" between the stars that nicely complements the book on *Stars* by the same author.

Melia, Fulvio. *The Black Hole at the Center of Our Galaxy*, Princeton, NJ: Princeton University Press, 2003. A contemporary account of the central 1,000 light-years of the Milky Way and the supermassive black hole that is thought to be dwelling there. Draws on recent radio, infrared, and X-ray observations of this mysterious region.

Payne-Gaposchkin, Cecilia. *Stars and Clusters*, Cambridge, MA: Harvard University Press, 1979. A stellar homage by one of the greatest astronomical pioneers of the twentieth century. Presents the characteristics and lives of stars through the insights gained from studying stellar clusters. Formed part of the Harvard Book Series on Astronomy.

Sasselov, Dimitar. *The Life of Super-Earths*, New York: Basic Books, 2012. A brief and lively introduction to the amazing discoveries that have revealed myriad exoplanetary systems in our solar neighborhood and beyond. The author provides an insider's perspective on the race to find planets via the transit method and to characterize these planets in terms of their internal structure and habitability. He makes the case that life is a planetary phenomenon, that super-Earths may fit the bill rather nicely, and that synthetic biology research on Earth may tell us what kinds of life-forms may exist out there.

Seeds, Michael A., and Dana Backman. *Horizons: Exploring the Universe*, 12th edition, Pacific Grove, CA: Brooks/Cole, Thomson Learning, 2012. One of about two dozen introductory astronomy textbooks currently available for nonscience majors. These full-color textbooks are continually updated with new material. They often feature planetarium software on CD and other learning tools online. This particular textbook explores the two central questions of "How do we know?" and "What are we?"

Stogon, Ronald, Stefan Binnewiess, and Susanne Friedrich. Transl. Klaus-Peter Schroeder. *Atlas of the Messier Objects: Highlights of the Deep Sky*, 2011. This beautiful book combines written descriptions, maps, and high-quality color images of the 110 sources that make up Charles Messier's famous compilation of "fuzzy" nonstellar objects. These include star clusters and nebulae in our own Milky Way Galaxy along with elliptical, lenticular, and spiral galaxies well beyond the Milky Way.

Waller, William H., and Paul W. Hodge. *Galaxies and the Cosmic Frontier*, Cambridge, MA: Harvard University Press, 2003. A descriptive introduction to the nature of galaxies and their lives over cosmic time—from their emergence shortly after the Hot Big Bang to their ongoing gyrations and transmutations. Includes chapters on the Milky Way, its neighbors, interacting and starbursting galaxies, quasars, and the overall structure and evolution of the universe. Complementary quantitative information is accessible at http://cosmos.phy.tufts.edu/cosmicfrontier.

Ward, Peter D., and Donald Brownlee. *Rare Earth—Why Complex Life Is Uncommon in the Universe*, New York: Copernicus–Springer-Verlag, 2000. Written by a paleogeologist and planetary astronomer, this book delineates the amazing set of circumstances that led to complex life on Earth. Besides providing a sobering antidote to untempered expectations for aliens everywhere, it offers well-grounded insights on planetary evolution, habitable zones, and prospects for microbial life beyond Earth.

Weissman, A. *The World Without Us*, New York: Thomas Dunne Books,

St. Michaels Press, 2007. A well-informed speculation upon what would happen if all humans on Earth suddenly perished. The author considers both short-term consequences (e.g., flooded subways) and longer-term developments (e.g., the evolution of plastic-eating microbes). Provides helpful perspectives on our temporal terrestrial legacy.

Wynn-Williams, Gareth. *The Fullness of Space*, Cambridge, UK: Cambridge University Press, 1993. An undergraduate-level introduction to the gaseous and dusty interstellar medium. Explains where the interstellar matter came from, what it consists of, how it collects together in clouds and clumps, and the way in which new stars and planets form from these nebular condensations.

TECHNICAL

Aller, Lawrence, H. *Atoms, Stars and Nebulae*, 3rd edition. Cambridge, UK: Cambridge University Press, 1991. Originally published in 1943, this undergraduate-level accounting of stellar and nebular astronomy has gone through three revisions. It is particularly strong in its description of stellar spectra and the analytic techniques that have been brought to bear for deriving physical properties of the stellar atmospheres and interiors.

Binney, James, and Michael Merrifield. *Galactic Astronomy*, Princeton, NJ: Princeton University Press, 1998. A comprehensive accounting of the structure, content, and kinematics of the Milky Way, along with complementary information on other nearby galaxies. The definitive textbook for upper-division undergraduates and beginning graduate students in astronomy.

Binney, James, and Scott Tremaine. *Galactic Dynamics*, Princeton, NJ: Princeton University Press, 1994. A theoretical treatment of the stellar and nebular motions in galaxies—including the Milky Way—and the gravitating agents that these motions manifest. Includes thorough explications of spiral density-wave theory, galaxy interactions, and the dynamical evidence for dark matter.

Burbidge, George, Allen Sandage, and Frank H. Shu, eds. *Annual Reviews of Astronomy and Astrophysics*, Palo Alto, CA: Annual Reviews, Inc. A yearly series of scholarly articles, each volume containing 10–20 reviews of particular research topics in astronomy and astrophysics, written by respected practitioners in their respective fields. Includes review articles on planetary, solar, stellar, nebular, galactic, extra-galactic, and cosmological topics.

Clemens, Dan, Ronak Y. Shah, and Tereasa Brainerd, editors. *Milky Way Surveys, The Structure and Evolution of Our Galaxy*, ASP Conference Series, Volume 317, San Francisco, CA: Astronomical Society of the Pacific, 2004. A collection of research reports on the Milky Way Galaxy, with a focus on recent observational surveys at multiple wavelengths.

Draine, Bruce T. *Physics of the Interstellar and Intergalactic Medium*, Princeton, NJ: Princeton University Press, 2010. A comprehensive graduate-level description of the physical processes associated with the gas and dust that occupy the space between stars within our Milky Way Galaxy and the far more tenuous material that fills the space between galaxies. Gaseous phases, mechanical and radiative energetics, and the cycling of gas from clouds to stars and back are fully described. The author is a leading expert on cosmic dust and gives a definitive treatment of this subject.

Matteucci, Francesca. *The Chemical Evolution of the Galaxy*, Dordrecht, the Netherlands: Kluwer Academic Publishers, 2001. A graduate-level review of galactic evolution with special focus on the stellar and interstellar processes that have enriched the Galaxy with ever greater abundances of heavier elements. Based on the research experiences and lecture notes of the author. Includes models of stellar nucleosynthesis and stellar outgassing, the contributions of novae and supernovae, and the effects of inflows and outflows. Presents estimates for the age of the Milky Way Galaxy and its younger disk component that are based on the measured abundances and known decay times of various radioactive isotopes.

Schulz, Norbert S. *From Dust to Stars*, Chichester, UK: Praxis Publishing–Springer, 2005. A graduate-level introduction to interstellar matter, molecular clouds and cores, concepts of stellar collapse, the evolution of young stellar objects, proto-planetary systems, and regions of clustered star formation.

Tielens, A.G.G.M. *The Physics and Chemistry of the Interstellar Medium*, Cambridge, UK: Cambridge University Press, 2005. A thorough graduate-level exposition of the interstellar medium, beginning with an introduction on "the galactic ecosystem" and launching into comprehensive treatments of all the ionic, atomic, molecular, and granular phases that permeate the ISM.

Vakoch, D. A. editor. *Communication with Extraterrestrial Intelligence*, New York: State University of New York Press, 2011. Proceedings of a NASA-sponsored astrobiology conference that focused on recent developments in the Search for Extraterrestrial Intelligence (SETI) and the emerging protocols for active Communication with Extraterrestrial Intelligence (CETI). Contains 32 articles by experts in this challenging and controversial field.

Zeilik, Michael, and Stephen A. Gregory. *Introductory Astronomy & Astrophysics*, 4th edition, Orlando, FL: Saunders College Publishing–Harcourt Brace, 1998. An upper-level undergraduate textbook that emphasizes the physics behind the astronomy of planets, stars, nebulae, galaxies, and the cosmos.

INDEX

• • • • • •